Russia's Hostile Measures in Europe

Understanding the Threat

Raphael S. Cohen, Andrew Radin

Prepared for the United States Army

For more information on this publication, visit www.rand.org/t/RR1793

Library of Congress Cataloging-in-Publication Data is available for this publication.

ISBN: 978-1-9774-0077-2

Published by the RAND Corporation, Santa Monica, Calif.

Preface

This report is the collaborative and equal effort of the coauthors, who are listed in alphabetical order. The report documents research and analysis conducted through 2017 as part of a project entitled *Russia, European Security, and "Measures Short of War,"* sponsored by the Office of the Deputy Chief of Staff, G-3/5/7, U.S. Army. The purpose of the project was to provide recommendations on how the Army can create options for the National Command Authorities to address the threat of Russian aggression in the form of measures short of war (or *hostile measures*) that leverage, improve upon, and develop new Army and joint capabilities.

The Project Unique Identification Code (PUIC) for the project that produced this document is RAN167301.

This research was conducted within RAND Arroyo Center's Strategy, Resources, and Doctrine Program. RAND Arroyo Center, part of the RAND Corporation, is a federally funded research and development center (FFRDC) sponsored by the United States Army.

RAND operates under a "Federal-Wide Assurance" (FWA00003425) and complies with the *Code of Federal Regulations for the Protection of Human Subjects Under United States Law* (45 CFR 46), also known as "the Common Rule," as well as with the implementation guidance set forth in DoD Instruction 3216.02. As applicable, this compliance includes reviews and approvals by RAND's Institutional Review Board (the Human Subjects Protection Committee) and by the U.S. Army. The views of sources utilized in this study are solely their own and do not represent the official policy or position of the U.S. Department of Defense or the U.S. government.

Contents

Figures and Tables

Figures

Tables

Summary

This report examines current Russian hostile measures in Europe and forecasts how Russia might threaten Europe over the near to medium term through activities that draw up short of the active use of lethal force. In line with George F. Kennan's use of the term measures *short of war*,[1] the term *hostile measures* encompasses a wide range of political, economic, diplomatic, intelligence, and military activities that could be perceived as harmful or hostile. The report assesses the range and limits of Russian tools of influence, European countries' abilities to resist or respond, and these states' vulnerabilities to hostile measures. The key finding in this report is that there appears to be an inverse relationship between Russia's means of influence and its ability to achieve its interests. Where it does have influence over actors, it tends not to be able to change policy significantly. This report examines three key regions of Europe: the Baltics; Southeastern Europe; and, briefly, the rest of Europe at large. For each region, we highlight Russian motives, opportunity, and means of influence.

What Are Hostile Measures?

Studying Russian hostile measures is inherently challenging because Russia does not want its sources and methods to be publicly known.

---

[1] George F. Kennan, in Giles D. Harlow and George C. Maerz, eds., *Measures Short of War: The George F. Kennan Lectures at the National War College, 1946–1947*, Washington, D.C.: National Defense University Press, 1991, pp. 3.

To address this challenge, we begin by specifying the evidence that would ideally be available to demonstrate Russian influence: evidence of intent from the Russian state, evidence of influence from Russia on a local proxy, and evidence that the Russian influence on the proxy led to the achievement of a Russian foreign policy objective. In practice, however, this level of evidence is almost never available. As a result, we cannot definitively prove the absence or existence of Russian influence; instead, we gather the available written and interview-based evidence and draw conclusions where possible.

Research by the RAND Corporation's Arroyo Center into Russian hostile measures determined that Russia pursues both directed and routine hostile measures. *Directed hostile measures* have a specific purpose or goal and appear to be employed as part of a cohesive plan; *routine hostile measures* are conducted by all arms of the Russian state as a matter of course. In this report, we identified five categories of very broad objectives that appear to guide Russian hostile measures in Europe:

1. pursuing security and survival of the regime
2. developing and maintaining great-power status
3. exerting influence within the *near abroad*, meaning Russia's immediate neighborhood and desired sphere of influence
4. increasing cooperation and trade with Western Europe
5. undermining enlargement of the European Union (EU) and North Atlantic Treaty Organization (NATO).

For the most part, Russia pursues broad, long-term objectives in Europe by applying some directed hostile measures but mostly routine ones. This "soft strategy" is applied with a wide range of different hostile measures. It is pursued with general long-term objectives in mind, but not necessarily in pursuit of specific goals. Consequently, we also observe a general Russian way of operating—increasing tension to create crisis and opportunity that Russia can then exploit (as opposed to preplanning how to achieve objectives through particular tactics).

The Baltics

The Baltic states of Estonia, Latvia, and Lithuania are often cited as being among the members of the EU and NATO that are most vulnerable to Russian influence. The Baltics are sometimes included in descriptions of Russia's near abroad, and Russia has exerted hostile measures in the region since the end of the Cold War. Possible motivations for Russian activities in the region include undermining and addressing the potential threat from the EU and NATO and maintaining longstanding connections with Russian speakers in the region. There are distinctions between Russian foreign policy interests in the Baltics and those in the other former Soviet states, however. For example, Russian analysts and some U.S. analysts of Russia note lower Russian interests and objectives for influence in the Baltics. Therefore, although Russia's intentions to use hostile measures in the Baltics remain real, the measures available and willingness to commit significant resources appear to be greater for other former Soviet republics, including Ukraine, Belarus, and the Central Asian states.

The Baltic states are mostly vulnerable to Russian hostile measures because of the demographic and economic legacies of being Soviet republics. Russia's most likely course of action is to attempt to leverage the Russian-speaking minorities in Estonia (30 percent) and Latvia (35 percent) that consume mainly Russian media. However, the integration of the Russian-speaking populations into Baltic culture and government structure means that such political subversion will be challenging. Russia could also use covert military force following modes of influence in Ukraine. However, given the intensive focus on internal security in the Baltic states, it will be difficult for Russian covert action to achieve a major change in the Baltic countries' alignment without support from overt forces that would, under Article 5 of the Washington Treaty, bring support from NATO countries. There is ongoing and valid concern about information operations, cyberattacks, and intimidation by Russian forces in the area; it appears that these measures of influence will likely continue to some degree for the foreseeable future. Finally, Russia could leverage economic or energy ties or other economic relations with the Baltics, but the Baltic states have taken measures to diversify their trade and dependence on Russia, and officials downplay the possible political impact in a cutoff in trade.

Baltic countries tend to prioritize the risks posed by conventional Russian military aggression in seeking assistance from their allies. Nevertheless, it remains important for the United States and other NATO allies to consider options to improve defense and deterrence against Russian hostile measures. Strengthening the conventional deterrent in the region could make all forms of Russian aggression less likely by reducing the potential for escalation, but there are risks of a large allied military presence in the region facilitating Russian hostile measures: Russia might take aggressive action because it perceives a threat, and local Russian speakers might oppose an allied military presence. One way to reduce the potential for Russian influence of the Russian-speaking population is to provide alternative Russian-language content and reduce the dominance of Russian-language media controlled by Moscow. U.S. bilateral engagement bolstering capabilities relevant to both conventional conflict and hostile measures—such as border security; internal and reserve security forces; and intelligence, surveillance, and reconnaissance capabilities—could also be especially beneficial.

Southeastern Europe

Southeastern Europe is a significant target of Russian influence, especially countries where a majority of the population speaks a Slavic language. Russian interest in Southeastern Europe is likely to arise in part from concern about growing NATO military capabilities, including ballistic missile defense capabilities in Romania; the goal of undermining EU and NATO enlargement in the Western Balkans and Moldova; a view that Russia should have influence over the region; and, perhaps most fundamentally, a desire to maintain Russian economic ties in the region.

Russia's historical, cultural, and economic ties with Southeastern Europe provide several opportunities for hostile measures. In Bulgaria, for example, Russian energy companies own key economic resources, and Russia funnels money to pro-Russian parties and groups. Russia has backed Serbia's position opposing Kosovo's independence and capitalized on a shared Orthodox religion and pan-Slavic beliefs to gain influence in the country. Russia also has sought to exploit ethnic and separatist conflict throughout the region, including through political support for Bosnian Serbs and military backing for the separatist region of Transnistria in Moldova. Russian military influence, through a developing anti-access and area denial capability in the Black Sea, is also cause for concern. Overall, Russian influence in Southeastern Europe is probably more dangerous to the stability, political progress, and economic development of the region than to the rest of Europe because Serbia and Moldova are not NATO or EU members, and Bulgaria is far too dependent on both institutions to openly challenge either of them.

Countering Russian hostile measures in Southeastern Europe will require whole-of-government engagement to develop domestic institutions to resist, identify, and counter Russian subversion. U.S. support will also be crucial to maintaining peace agreements in the Western Balkans. Finally, U.S. military assistance could relieve some of the continued pressure on the region stemming from migration and terrorism—and, in the process, mitigate potential points of vulnerability to Russian hostile measures.

Other Regions

Turning to the rest of Europe, major U.S. allies in Western Europe present the largest strategic prize to Russia outside of the United States. Western Europe is home to Europe's largest economies, most-powerful militaries, and key U.S. military bases. Still, this region seems comparatively less vulnerable to Russian influence, although Russia might attempt to use a variety of hostile measures against it. It is true that certain Western senior politicians have financial interests in Russia—potentially opening them to Russian influence—but most are former policymakers rather than ones who are currently serving. Moreover, as of early 2018, there was no documented public example where Russia used these ties to change policy decisions. Still, even if Russia cannot directly shift policy, it might have an opportunity to fuel confusion and dissension in these countries.

Russia also maintains economic and energy ties throughout Europe. But Europe continues to make strides in energy diversification, and its economic ties are also comparatively small when measured against other trading partners. Russia has an information operation effort in Western Europe through such media as RT (formerly called Russia Today), but its viewership pales in comparison with other media outlets. The Russian military also periodically conducts air and naval shows of force targeted at Western Europe, but these displays of force have so far prompted increased NATO action, not intimidation.

There are two key caveats to this relatively rosy assessment of much of Europe's ability to withstand Russian hostile measures. First, smaller Central European states are, by and large, more vulnerable to Russian hostile measures than their larger Western European neighbors. These countries are poorer and less developed democratically, and some have significant Slavic populations that are more favorably disposed to Russia. At the same time, these countries are also less influential and pose less of a potential prize for Russia than their Western peers. Second (and, perhaps, more ominously), Russia has cultivated relationships with extremist parties throughout Europe, particularly on the far right—and, for reasons quite apart from Russian actions, these parties are on the rise. Given Europe's ongoing challenges in the areas of economics, terrorism, and immigration, these extremist parties will pose a significant vulnerability for Europe over the near term.

Conclusion

The precise policy prescriptions vary by region but, in all cases, countering Russian hostile measures requires a response by European countries supported by the whole of the U.S. government, not just the Joint Force and the U.S. Army. A comprehensive approach must include strengthening the rule of law to address co-opting of criminal enterprises, strengthening democratic institutions to counter pro-Russian parties, and development of alternative media sources to counter Russian information operations. The Joint Force and the U.S. Army have a lead role in only a few of the efforts to respond to hostile measures. Nevertheless, the U.S. military is often one of the most capable U.S. government organizations present and frequently has unique capabilities it can contribute.

We also highlight three overarching lessons for the Joint Force at large and the U.S. Army. First, in deploying forces to Europe to counter Russian aggression, the Army should also prepare to defend against and counter Russian hostile measures. The Joint Force and U.S. Army must also consider how Russia might respond aggressively to any forward-deployed forces. Second, the Army should develop counterintelligence, public affairs, civil affairs, and other key enablers to better counter Russian hostile measures. Finally, responding to Russian hostile measures places a new premium on political awareness, as well as on crisis management. U.S. military personnel need to be aware of Russian hostile measures—particularly when deployed in countries with frozen conflicts or where there is a large pro-Russian population—to help avoid accidentally sparking a crisis. Whatever the U.S. response, preparation for involvement in a wide range of conflicts can help reduce the risk of mismanagement, miscalculation, and escalation.

Acknowledgments

We would like to thank several individuals who made this report possible. First, we would like to thank Major General William Hix and Mr. Walter Vanderbeek for funding and supervising this project. We also would like to thank Ben Connable, the project leader, and Sally Sleeper, RAND Arroyo Center's Strategy Doctrine and Resources program director, for providing the funding and guidance for this research. Additionally, Catherine Dale of RAND, and Bruce McClintock of Zenith Advisors Group, offered thoughtful comments on a previous draft and improved the final product tremendously. We would like to thank the presenters and participants at a RAND symposium on Russian Measures Short of War, held in Cambridge, United Kingdom, in February 2016. Additionally, we would like to thank the staffs of the U.S. embassies in Sofia, Bulgaria, and in Bucharest, Romania, for facilitating a research trip to both countries in June 2016. The U.S. embassies in the Czech Republic, Estonia, Hungary, Latvia, Lithuania, and Slovakia also facilitated other research trips in previous years. Above all, however, we owe a special debt of gratitude to the dozens of government officials, military officers, think tank analysts, and academicians—drawn from across all the aforementioned countries in Europe—who gave their time and insights to help shape this report. Although human subjects protocols prevent us from listing everyone's name here, this report would not have been possible without their time and expertise.

Abbreviations

AfD	*Alternative für Deutschland* (Alternative for Germany)
ATAKA	Attack Party
BMD	ballistic missile defense
DoD	U.S. Department of Defense
EU	European Union
FPÖ	Austria Freedom Party
FSB	Federal Security Service
GDP	gross domestic product
ISR	intelligence, surveillance, and reconnaissance
KAPO	Estonian Internal Security Service
KGB	*Komitet Gosudarstvennoy Bezopasnosti* (Soviet Secret Service)
LNG	liquefied natural gas
MSW	Russian Measures Short of War
N/A	not applicable
NATO	North Atlantic Treaty Organization
NGO	nongovernmental organization
OHR	Office of the High Representative
RS	*Republika Srpska* (Serb Republic)
RT	news outlet formerly known as Russia Today
UKIP	United Kingdom Independence Party
UN	United Nations
WoG	whole of government

CHAPTER ONE

Introduction

This report examines recent Russian hostile measures in Europe and forecasts how Russia might threaten Europe through activities short of the active use of lethal force over the near to medium term. Our use of the term *hostile measures* encompasses a wide range of political, economic, diplomatic, intelligence, and military activities that could be perceived as harmful or hostile. We use it synonymously with the term *measures short of war*, which was originally used by George F. Kennan to describe a wide range of Soviet activity that stopped short of actual warfare and was intended to pursue Soviet interests on a routine basis.[1] Because Russia would presumably perform these actions during periods of war and of peace, however, we prefer the term hostile measures.

In this report, we assess the range and limits of Russian tools of influence, the ability of European countries to resist or respond, and the resulting major vulnerabilities of European states. One conclusion is that there is an inverse relationship between Russia's means of influence over an organization or notable individuals within a given country and the degree to which Russia can use that country to achieve its larger strategic interests. We also observe a general Russian way of operating—increasing tension to create crisis and opportunity that Russian can then exploit (as opposed to preplanning how to achieve objectives through particular tactics).

---

[1] George F. Kennan, in Giles D. Harlow and George C. Maerz, eds., *Measures Short of War: The George F. Kennan Lectures at the National War College, 1946–1947*, Washington, D.C.: National Defense University Press, 1991, pp. 3.

Our analysis focuses on Russian hostile measures in three key regions: the Baltics; Southeastern Europe (both commonly cited as potential key battlegrounds for Russian); and, briefly, the rest of Europe. Accounts of the Soviet Union's use of hostile measures during the Cold War and Russia's use of these tactics in recent crises in former Soviet republics, including Ukraine and Georgia, sometimes focus on military tools of influence, such as the use of special operations forces. Although there is some risk of these tactics being used in the Baltics (and perhaps in Southeastern Europe), a likelier threat is posed by Russia's nonlethal tool kit—including corruption, funding of political proxies, economic leverage, information operations, and exploitation of Russian exiles. Possible effects include influencing the discourse within the target society about Russia and Europe, creating internal instability or conflict, or shaping policy within the target country on key issues. Assessing the nature of this risk and identifying possible responses is critical for framing U.S. foreign policy in Europe.

Organization of This Report

In the remainder of this chapter, we discuss the study's methodology, noting the significant challenge of concretely identifying and tracing covert and criminal activities. Then, we consider Russian objectives for pursuing hostile measures in Europe based on Russia's core foreign policy interests. Third, we analyze Russia's overall strategy to use hostile measures in Europe in the near future; specifically, we expect Russia to pursue a wide range of low-cost and low-risk tactics across many European countries simultaneously to achieve its goals, without specific expectations that any given tactic will work. Finally, we offer a three-dimensional typology of proxy groups that describes Russia's influence on groups, groups' alignment with Russian interest, and groups' ability to influence policy in the target society.

We then discuss Russian hostile measures in three regions. In Chapters Two and Three, we focus on the Baltics (Estonia, Latvia, and Lithuania) and Southeastern Europe (the former Yugoslavia, Romania, Bulgaria, Moldova, and Greece) as the areas that are potentially most susceptible to Russian hostile measures. In Chapter Four, we broaden the aperture and look at the potential for Russia to conduct hostile measures in Central and Western Europe, including key U.S. allies, such as the United Kingdom, France, and Germany. Rather than using each chapter to catalogue potential use of each type of hostile measure in each country, we highlight the main Russian motives, opportunities, and means of using hostile measures in each region.

A final section, drawing from the empirical chapters, identifies the likelihood of hostile measures occurring and the severity of results if they were to be effective. From this analysis, we rank and prioritize the threats in different European countries. We also identify policy implications for the U.S. Army and other U.S. and European government institutions to defend Europe from Russian hostile measures.

Methodology

Studying Russian hostile measures is inherently challenging. Many of these measures are covert, criminal, or denied. Even if Russia is pursuing such measures, it often does not want its sources and methods to be publicly known. To address this challenge, it is useful to think hypothetically of what evidence would ideally be available to demonstrate Russian influence and aggression. As shown in Figure 1.1, with any given hostile measures, there should be evidence of intent and influence from someone within or controlled by the Russian state, evidence of the influence from Russia acting on a local proxy, and evidence that influence from Russia has led to the achievement of a given Russian foreign policy objective.

In practice, this level of evidence is almost never available. If evidence of Russia's connections with a local proxy is weak or absent (perhaps because the link between the government and a state-influenced actor is missing), this could mean that the local proxy is taking action that is friendly to Russia but not because of Russian state action. If there is evidence that Russia has influence over a group but it is unclear how this group's activities benefit Russia, it might be that Russia's activities are benign. In the absence of evidence, we do not

Figure 1.1
Examples of Desired Evidence of Russian Hostile Measures

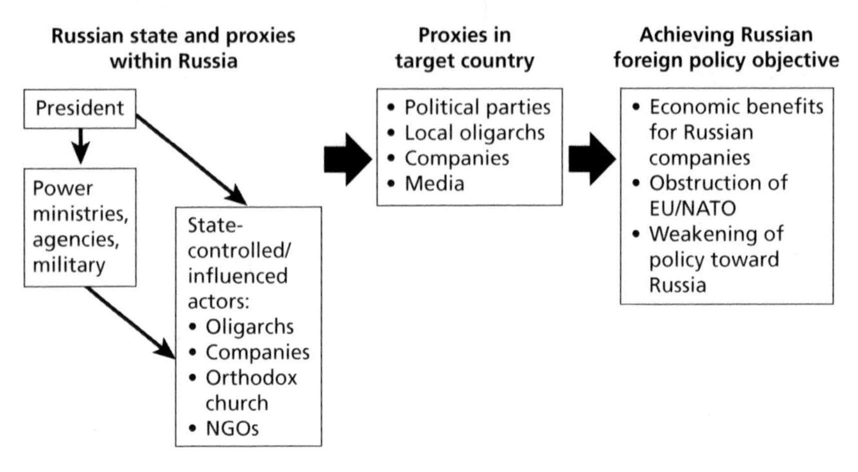

NOTE: EU = European Union; NATO = North Atlantic Treaty Organization; NGO = nongovernmental organization.
RAND *RR1793-1.1*

make assumptions but instead seek to state the available evidence and draw possible conclusions.

To gather evidence of Russian influence, we draw on scholarly work, media accounts, and other open sources; interviews with European defense and security analysts; and discussions from a RAND symposium on Russian Measures Short of War (MSW) hosted in Churchill College, Cambridge University, in the United Kingdom. Because of the human subjects protocol for this and other projects, our discussions were conducted on a not-for-attribution basis.

We used these discussions to identify the range of possible and relevant Russian hostile measures in Europe. In identifying a scenario or potential means of Russian influence, we are not assessing the likelihood or potential for it to occur; we are simply identifying it for consideration. In the sections that follow, we analyze these means or scenarios based on our analysis of the environment and Russian capability, then evaluate the threat and necessary response.

Two additional caveats are in order. First, because this study looks at future potential uses of Russian hostile measures, it is fundamentally predictive and thus more speculative than historical studies of past or even current hostile measures. Second, although every effort is made to document assertions, this research is necessarily more anecdotal and circumstantial than historical studies are.

To organize the analysis, each of the three regional analyses is divided into three sections: motives, opportunity, and means. In discussing motives, we consider Russia's potential motivations for pursuing hostile measures in the region, drawing from both general Russian objectives and hypothesized motivations identified by our interlocutors. For example, in Western Europe, Russia seeks respect as a great power, but it also seeks to influence overall EU and NATO policy so that its regional interests are not threatened. In discussing opportunity, we outline the general political, social, and economic factors that make the relevant countries vulnerable to Russian hostile measures and the countervailing strengths that might make them less vulnerable. In the sections on means, we identify and assess the major Russian hostile measures that might be applied to the region, drawing from Russia's policies, organizations, and past behavior in comparable contexts, and from preparations in these countries to counter Russian hostile measures.

Russian Objectives

This section outlines five general goals of Russian foreign policy that would most likely lead Russia to pursue hostile measures in Europe in the next five years.[2]

First and foremost, Russia pursues its own security and the preservation of the regime. Analysts highlight two factors to explain Russia's imperial ideology and persistent fear of outside invasion: Russia's geographic position, which lacks major natural barriers, and its his-

---

[2] This section draws from official documents, analysis, and interviews detailed in Andrew Radin and Clinton Bruce Reach, *Russian Views of the International Order*, Santa Monica, Calif.: RAND Corporation, RR-1826-OSD, 2017, pp. 7–23.

tory of foreign invasion.[3] Stephen Kotkin, a Princeton professor, writes, "Russia has felt perennially vulnerable and has often displayed a kind of defensive aggressiveness. . . . Today, too, smaller countries on Russia's borders are viewed less as potential friends than as potential beachheads for enemies."[4] During the Soviet period, Russian leaders repressed domestic opposition, justifying their actions based on the fear of external attack. Since the end of the Soviet Union, Russian leaders have continued to fear domestic upheaval and popular protest, especially following so-called color revolutions in Georgia and Ukraine. Gleb Pavlovsky, a former adviser to Russian leader Vladimir Putin, notes "a feeling of great vulnerability" in the Kremlin since 1993, "an absolute conviction that as soon as the power centre [sic] shifts, or if there is mass pressure, or the appearance of a popular leader, then everybody will be annihilated."[5] Some analysts hypothesize that Russia's foreign policy might be intended to bolster the position of the regime at home. There was a substantial increase in the popularity of the Putin regime during crises in Georgia and Ukraine in 2008 and 2014, respectively, and theoretically the regime might be able to use its foreign policy to increase its popularity at home.[6]

For many Russian officials and analysts, Russia's policy in the information sphere is a key part of its national security. Chatham House Russia expert Keir Giles, for example, notes that Russia "considers itself to be engaged in full-scale information warfare, involving not

---

[3] Jeffrey Mankoff, *Russian Foreign Policy: The Return of Great Power Politics*, Lanham, Md.: Rowman and Littlefield, 2012, p. 3. Kennan, for example, writes in his "Long Telegram," "At the bottom of Kremlin's neurotic view of world affairs is a traditional and instinctive Russian sense of insecurity. Originally, this was insecurity of a peaceful agricultural people trying to live on vast exposed plain [sic] in neighborhood [sic] of fierce nomadic peoples." George F. Kennan, "Long Telegram" (Moscow to Washington), National Security Archive, February 22, 1946.

[4] Stephen Kotkin, "Russia's Perpetual Geopolitics: Putin Returns to Historical Patterns," *Foreign Affairs*, May/June 2016.

[5] Gleb Pavlovsky, "Putin's World Outlook," *New Left Review*, No. 88, July–August 2014, p. 62.

[6] Olga Oliker, Keith Crane, Lowell H. Schwartz, and Catherine Yusupov, *Russian Foreign Policy: Sources and Implications*, Santa Monica, Calif.: RAND Corporation, MG-768-AF, 2009, pp. xiv–xvii, 43–44; Sam Greene and Graeme Robertson, "Explaining Putin's Popularity: Rallying Round the Russian Flag," *Washington Post*, September 9, 2014.

only offensive but also defensive operations," with the latter focused on protecting the "national information space" of Russia from foreign intrusion.[7] Russia also might use offensive information warfare, such as support for propaganda, with the defensive intent of maintaining the regime's own control over information internally.[8]

Second, Russia sees itself as a great power and seeks recognition as one of the world's great powers. Jeffrey Mankoff, Dmitri Trenin, and others highlight Russia's identity as a great power as a consistent and dominant feature of Russian foreign policy.[9] Russia, from this point of view, is one of the major powers within a multipolar world, and should be treated with respect appropriate to this status—including a seat at the table in resolving major international crises and veto power in such major institutions as the United Nations (UN)[10] —and not, as Fyodor Lukyanov puts it, as simply another "big Poland" in Eastern Europe.[11]

---

[7] Keir Giles, *Russia's 'New' Tools for Confronting the West Continuity and Innovation in Moscow's Exercise of Power*, London: Chatham House, March 2016, pp. 27, 33.

[8] See Timothy H. Thomas, "Russia's Information Warfare Strategy: Can the Nation Cope in Future Conflicts?" *Journal of Slavic Military Studies*, Vol. 27, No. 1, 2014; Jolanta Darczewska, *The Anatomy of Russian Information Warfare: The Crimean Operation, a Case Study*, Warsaw, Poland: Centre for Eastern Studies, May 2014.

[9] Mankoff, 2012, p. 12. Dmitri Trenin writes,

> When he became foreign minister in 1996, Yevgeny Primakov famously proclaimed: "Russia has been, is, and will be a great power!" This became a rallying point for the Russian elite. Virtually everyone chimed in. What was less clear was what it meant to be a "great power" in the new era.

Dmitri Trenin, *Post-Imperium: A Eurasian Story*, Washington, D.C.: Carnegie Endowment for International Peace, 2011, p. 205. See also Lilia Shevtsova, *Russia: Lost in Transition*, Washington, D.C.: Carnegie Endowment for International Peace, 2007, p. 3.

[10] Of Russian behavior in Bosnia and Herzegovina, for example, Richard Holbrooke writes that

> we felt that Moscow's primary goal was neither to run nor to wreck the negotiations. Rather, what it wanted most was to restore a sense, however symbolic, that they still mattered in the world.

Richard Holbrooke, *To End a War*, New York: Modern Library, 1999, p. 117.

[11] Lukyanov writes,

> For all practical intents and purposes, a large country with the mentality and history of an independent great power simply could not overnight turn itself into a 'big Poland'

There are a wide range of hostile measures that Russia could use to bolster its influence, prestige, and autonomy along these lines.

Third, Russia pursues influence within its neighborhood. This desire likely arises both from its self-conception as a great power and from deeply rooted ideas in in Russian society. A 2009 RAND report explains that the reasons for Russia's interest in its near abroad "stem from Russia's quest for prestige, its history, its economic priorities, and its fundamental security concerns."[12] Russian analysts note an "imperial" identity—an idea that Russian interests extend beyond Russia's borders.[13] At a minimum, this identity implies a greater degree of autonomy, and exclusive influence and control within its region.[14] Indeed, in a 2016 survey of Russian elites, 82.3 percent responded that the national interests of Russia "for the most part extend beyond its existing territory," up from 43.3 percent in 2012 and 64 percent in 2008.[15]

> and follow in the footsteps of states seeking admission to the EU and NATO—institutions that, in any event, never offered membership to Russia.

Fyodor Lukyanov, "The Lost Twenty-Five Years," *Russia in Global Affairs*, February 28, 2016.

[12] Oliker, Crane, et al., 2009, p. 93.

[13] Igor Zevelev, former head of the MacArthur Foundation office in Moscow, notes that Russian identity includes "Little Russians" (Ukrainians), "White Russians" (Byelorussians), and "Great Russians" (ethnic Russians). Igor Zevelev, *NATO's Enlargement and Russian Perceptions of Eurasian Political Frontiers*, Garmisch-Partenkirchen, Germany: George Marshall European Center for Security Studies, NATO, undated, p. 17.

[14] Russian President Dmitry Medvedev explained,

> The world should be multipolar. Unipolarity is unacceptable; domination is impermissible. We cannot accept a world order in which all decisions are taken by one country, even such a serious and authoritative country as the United States of America. This kind of world is unstable and fraught with conflict. Russia, just like other countries in the world, has regions where it has its privileged interests. In these regions, there are countries with which we have traditionally had friendly cordial relations, historically special relations. We will work very attentively in these regions and develop these friendly relations with these states, with our close neighbours.

Paul Reynolds, "New Russian World Order, The Five Principles," BBC News, September 1, 2008.

[15] Sharon Werning Rivera, James Byan, Brisa Camacho-Lovell, Carlos Fineman, Nora Klemmer, and Emma Raynor, *The Russian Elite 2016: Perspectives on Foreign and Domestic Policy*,

A major question, then, is how much influence Russia seeks to exert over which countries in Europe. Some analysts regard the entire former Soviet Union as Russia's imagined near abroad and thus its primary desired sphere of influence—indeed, this view appears in analysis of Russian intentions from within the Baltic states.[16] Some Russian analysts, in contrast, describe the geographic extent of the near abroad as the former Soviet countries excluding the Baltic states, which Russia has abandoned to the Western sphere of influence.[17] Some more-radical views within Russia support greater expansion. Aleksandr Dugin, a prominent Eurasianist, for example, argues for the creation of a Eurasian Union, which would be an "analogue of the USSR [Soviet Union] on a new ideological, economic and administrative basis." Poland, Latvia, and Lithuania, among other countries, would be included in this union with a "special status."[18] Although Dugin is representative of more-radical views of Russian imperial thought, neither he nor others sharing a similar ideology appear especially influential to the regime.[19] Rather than lumping together all former Soviet countries in a single category, a more-precise approach might be to differentiate between

---

Clinton, N.Y.: Arthur Levitt Public Affairs Center, Hamilton College, May 11, 2016, p. 15.

[16] See Gatis Pelnens, ed., *The 'Humanitarian Dimension' of Russian Foreign Policy Toward Georgia, Moldova, Ukraine, and the Baltic States*, Riga, Latvia: Centre for East European Policy Studies, International Centre for Defence Studies, Centre for Geopolitical Studies, School for Policy Analysis at the National University of Kyiv-Mohyla Academy, Foreign Policy Association of Moldova, International Centre for Geopolitical Studies, 2009, p. 18.

[17] Discussions with U.S. and Russian think tank analysts, April 2016, Washington D.C.; Marlene Laruelle, *The 'Russian World': Russia's Soft Power and Geopolitical Imagination*, Washington, D.C.: Center on Global Interests, May 2015, p. 1; Trenin, 2011, p. 107.

[18] Estonia would be within Germany's sphere of influence under his conception. John B. Dunlop, "Aleksandr Dugin's Foundation of Geopolitics," *Demokratizatsiya*, Vol. 12, No. 1, January 31, 2004.

[19] These thinkers have been highlighted in accounts analyzing Russia's motivation and strategy in Ukraine, although without specific evidence that their ideas were influential in strategic planning. See Robert R. Leonhard and Stephen P. Phillips, *'Little Green Men': A Primer on Modern Russian Unconventional Warfare*, Fort Bragg, N.C.: U.S. Army Special Operations Command, Assessing Revolutionary and Insurgent Strategies Project, undated, pp. 15–17; Darczewska, 2014; Marlene Laruelle, *Russian Eurasianism: Ideology of Empire*, Washington, D.C.: Woodrow Wilson Center Press, 2012, p. 113.

Russian goals of influence toward different sets of countries. Indeed, Russian strategic documents and its policies in the former Soviet Union indicate that Russia has a more active policy in the non-Baltic former Soviet countries (including Central Asia, Belarus, Ukraine, Moldova, Georgia, Armenia, and Azerbaijan) than in the Baltics or other EU and NATO members.[20] A 2017 RAND report identifies different spheres of influence—from Russia, Belarus, and Central Asia outward—in which Russia has a decreasing desire for influence.[21]

Fourth, Russia seeks economic prosperity, which in turn, likely requires some level of cooperation, trade, and investment with Europe. Boris Yeltsin's government initiated a process of integrating Russia with Western economic institutions in the 1990s.[22] Putin and Medvedev maintained the goal of trading with the West and joining such Western-led economic institutions as the World Trade Organization even while

---

[20] Trenin, for example, writes,

> When in 2003 Russia redeployed forces from the Balkans and "conceded" the Baltics—under Putin, unlike in the Yeltsin period, there was no vociferous campaign protesting their membership, just clenched teeth— this regrouping was done to better consolidate Russia's few assets where it mattered most: in the CIS [Commonwealth of Independent States]. Moscow was ready to renounce its claim on a role in its old sphere of interest: Central and Southeastern Europe, and the Baltics. But it resolved not to allow further Western encroachments into the territory it felt was its "historical space."

Trenin, 2011, p. 107. See also Russia's framing of its own regional interests in its 2013 Foreign Policy Concept. Russian Federation, Ministry of Foreign Affairs, Concept of the Foreign Policy of the Russian Federation, Section IV, February 12, 2013.

[21] Radin and Reach, 2017, p. 11.

[22] The 1993 Russian Foreign Policy Concept, for example, stated that "Russia will strive toward the stable development of relations with the United States, with a view toward strategic partnership and, in the future, toward alliance." See Interdepartmental Foreign Policy Commission of the Security Council, Russian Federation, "Russian Foreign Policy Concept," April 1993, quoted in Speaker's Advisory Group on Russia, U.S. House of Representatives, "From Friendship to Cold Peace: The Decline of the U.S. Russia Relations During the 1990s," *Russia's Road to Corruption: How the Clinton Administration Exported Government Instead of Free Enterprise and Failed the Russian People*, Washington, D.C.: U.S. Congress, via Federation of American Scientists website, September 2000.

they adopted increasingly harsh rhetoric toward the West.[23] At a speech at Munich in 2007 that is frequently cited as showing Russia's increasing opposition to the West, Putin emphasized the "openness and stability of the Russian economy," and Russia's pursuit of joining the World Trade Organization.[24] Russia's economic investment, trade, and linkages with Europe are fundamental means of increasing Russia's economic prosperity and serve as a potential form of leverage for Russian foreign policy, as will be discussed. Still, there are numerous examples of Russia compromising on its economic interests for political reasons. For example, Russia's goal of being recognized as a great power has undermined its efforts at negotiation with the EU,[25] and Russian companies have apparently based natural gas prices in part on geopolitical priorities.[26]

Fifth, and related to many of the aforementioned objectives, Russia seeks to stop EU and NATO enlargement and to undermine EU and NATO activities in Russia's perceived sphere of influence. Russia has expressed significant concern about NATO enlargement. In 2007, Putin described NATO enlargement as "a serious provocation that reduces the level of mutual trust" and expressed concern about NATO bases on Russia's border.[27] Similarly, in 2016, Russia's Foreign Minister Sergey Lavrov wrote that the choice to pursue NATO enlargement "is the essence of the systemic problems that have soured Russia's rela-

---

[23] Mankoff describes Russian foreign policy from 2000 to 2004 as having a

> generally passive, reactive nature—a development that led some analysts to argue that a historic turning point had been reached, bringing an end of the era of confrontation between Russia and the West once and for all.

Mankoff, 2011, pp. 30–33.

[24] Vladimir Putin, "Speech and the Following Discussion at the Munich Conference on Security Policy," Munich, Germany, February 12, 2007.

[25] Carl Bildt, *Russia, the European Union, and the Eastern Partnership*, Latvia: European Council on Foreign Relations, ECFR Riga Series, May 19, 2015.

[26] Rawi Abdelal, "The Profits of Power: Commerce and Realpolitik in Eurasia," *Review of International Political Economy*, Vol. 20, No. 3, June 2013.

[27] "Putin's Prepared Remarks at 43rd Munich Conference on Security Policy," transcript via *Washington Post*, February 12, 2007.

tions with the United States and the European Union."[28] Russia's military leadership is similarly concerned that NATO's enlargement brings adversary forces closer to the country,[29] and Russia's 2016 National Security Strategy highlights the threat of NATO.[30] Although Russia was generally supportive of the EU until 2013, its strong opposition to Ukraine's steps toward integration with Europe shows increasing concern about EU enlargement. Putin, for example, argued that the EU's integration effort with Ukraine was mistakenly intended "to disrupt an attempt to re-create the Soviet Union."[31] Deputy Prime Minister Dmitri Rogozin also connected Ukraine's decision to sign the Association Agreement with eventual NATO membership.[32] Some argue that Russia seeks to undermine the entire Western system, including destroying NATO, but this objective is contested. Russian strategic documents focus on the enlargement of NATO and analysts highlight

---

[28] Sergey Lavrov, "Russia's Foreign Policy: Historical Background," *Russia in Global Affairs*, via Russian Federation Ministry of Foreign Affairs website, March 3, 2016.

[29] Dmitri Gorenburg summarizes a 2015 speech by Chief of the General Staff Valery Gerasimov:

> The most significant threat facing Russia, in Gerasimov's view, comes from NATO. He highlights the threat from NATO enlargement to the east, noting that all 12 new members added since 1999 were formerly either members of the Warsaw Pact or Soviet republics. This process is continuing, with the potential future inclusion of former Yugoslav republics and continuing talk of perspective Euroatlantic integration of Ukraine and Georgia.

Dmitri Gorenburg, "Moscow Conference on International Security 2015, Part 2: Gerasimov on Military Threats Facing Russia," *Russian Military Reform* blog, May 4, 2015.

[30] Specifically, the National Security Strategy says,

> The buildup of the military potential of the North Atlantic Treaty Organization (NATO) and the endowment of it with global functions pursued in violation of the norms of international law, the galvanization of the bloc countries' military activity, the further expansion of the alliance, and the location of its military infrastructure closer to Russian borders are creating a threat to national security.

Russian Federation, National Security Strategy, Moscow, 2016.

[31] Vladimir Soloyov, "*Miroporyadok* [World Order]," YouTube, streamed live December 20, 2015.

[32] Sergey Aleksashenko, "For Ukraine, Moldova, and Georgia, Free Trade with Europe and Russia Is Possible," Carnegie Middle East Center, July 3, 2014.

that the destruction of NATO could be destabilizing and threatening to Russian interests.[33]

Russian Strategies

Russia likely intends to use hostile measures—a range of tools of statecraft—to achieve these objectives. In pursuing these tactics, Russia does not appear to have in mind a clear causal logic about how to achieve its objectives. Instead, it appears to apply many different measures of influence simultaneously in pursuit of general long-term objectives. Russian leaders likely hope that these measures have a general impact in favor of their objectives (for example, weakening NATO cohesion), or that these measures create tension and lead to crises that Russia can later exploit. Hence, although Russia does have agency in creating and utilizing tension in Europe, it operates in an opportunistic fashion, exploiting circumstances, as they arise, to achieve its strategic ends.

We refer to this approach as a *soft strategy*. It differs from more linear approaches that clearly specify intermediate goals and expectations for the outcomes of specific actions. For example, Russia does not expect that support for a particular far-right party in France or Hungary will accomplish the goal of undermining NATO and the EU; nor does it expect that this support will lead to the election of a particular leader who will accomplish that goal. Rather, there is a hope that support for the far-right party will, in combination with other activities or through an unexpected or fortuitous set of occurrences, eventually bring about an opportunity or circumstance that Russia can exploit. Because it is unlikely that any given strategy or tactic is most likely to succeed, and because these tactics might not achieve Russia's objectives even if they are successful, Russia appears to be pursuing a wide range of such tactics at the same time.

Although there is no specific proof that Russian leaders have adopted this approach, we find this strategic perspective convincing

---

[33] Andrew Radin, "How NATO Could Accidentally Trigger a War with Russia," *The National Interest*, November 11, 2017b.

for several reasons. First, it has been reiterated in a wide range of discussions with senior officials and analysts in the United States, Baltics, and Southeastern Europe.[34] Second, our observations of Russia's past actions and possible means of influence in the future reflect this strategic perspective. In each region, there does not appear to be a single dominant means of influence; rather, Russia appears to be taking—and have the potential to take—a range of possible actions that in combination might achieve its ultimate goals.

The account of Russia's soft strategy is supported by other analysis. Scott Morrison, former director of the Commander's Action Group for NATO Special Operations Headquarters, writes that Russia's approach is better understood using "systemic operational design" (SOD),[35] an Israeli concept focusing on "the relationships between entities within a system to develop rationale for systemic behaviors that accounts for the logic of the system, facilitating a cycle of design, plan, act, and learn."[36] Unlike conventional Western approaches to strategic planning, where there are clear expectations about the effect that will result from a particular action, SOD involves various discourses and evolving discussions to develop military planning. For example, "[t]he SOD approach aims at recognizing a range of actions and expresses tensions within or between entities," with the goal that actions or exploitation of tensions might lead to the achievement of underlying goals.[37]

Similarly, descriptions of Putin's strategic logic align with our account of Russia's soft strategy. Sergei Pugachev, a Russian businessman who knew Putin before he became president, noted that "Putin is not someone who sets strategic plans; he lives today." Pugachev

[34] In Romania, for example, one analyst responded to a question about Russian goals by saying that there was no specific immediate objective, but that Russia was "seeking an opportunity." Similarly, a senior Romanian military officer noted that Russia "will use the full range of hybrid tactics" to regain its position. Interviews with Romanian analysts and senior Romanian officer, Bucharest, June 23–24, 2016.

[35] We thank Morrison for sharing his draft manuscript with us.

[36] William T. Sorrells, Glen R. Downing, Paul J. Blakesley, David W. Pendall, Jason K. Walk, and Richard D. Wallwork, *Systemic Operational Design: An Introduction*, Fort Leavenworth, Kan.: School of Advanced Military Studies, AY 04-05, May 26, 2005, p. i.

[37] Sorrells et al., 2005, p. 18.

further notes that Putin regularly arrived at work with only "well-sharpened pencils, a clean sheet of paper and a newspaper. . . . There were no documents, nothing."[38] Fiona Hill draws from Putin's background in intelligence to understand his approach, observing "Putin *is* a strategist—if we understand what that term means for someone coming from a background in the Soviet-era secret service called the *Komitet Gosudarstvennoy Bezopasnosti* (KGB). For Putin, to plan strategically means planning for contingencies. You have to expect the unexpected, be able to learn from mistakes (both your own and those of others), and to adapt. . . . You have to keep your options open and have backup plans."[39] Other U.S. and Russian analysts describe Putin as a tactician, responding to day-to-day events without a detailed strategy, but with long-term objectives but in mind.[40]

There are several implications of Russia's soft approach to strategy. First, the observation of a particular activity by Russia is not necessarily indicative of a particular strategy for achieving its objectives. Support for the Front National in France does not indicate a specific goal of overturning the government of France and thereby gaining control of NATO policymaking. U.S. and NATO officials should not assume that Russia's adoption of a tactic means that it will pursue subsequent activities that appear logically connected. Second, a particular Russian tactic need not be associated with only one of its underlying objectives. For example, Russian attempts to gain influence in Bulgaria could play equally to Russia's objective of expanding its economic influence and to undermining the EU and NATO by encouraging Bulgaria to break consensus on core policy issues.

---

[38] Pugachev further observed,

> I had been in politics about ten years and seen everyone. They'd have tons of documents. They'd always be doing something. But with him it was just quiet, no one there, no meetings, everything quiet. He'd sit there, or watch TV. He really likes watching TV.

Oliver Bullough, "Former Aide Says Putin Has No Strategic Plans," *Time*, November 5, 2014.

[39] Fiona Hill, "Putin: The One-Man Show the West Doesn't Understand," *Bulletin of the Atomic Scientists*, Vol. 72, No. 3, 2016, p. 2.

[40] Discussion with U.S. and Russian analysts, Washington D.C., March and April 2016.

Dimensions of Russian Influence

Russian influence over groups or individuals within a country is a fundamental element of its use of hostile measures. However, individuals and groups could fall under a range of several types or forms of influence—some individuals might directly follow Russia's orders; others might share similar interests on a given topic. We identify three different dimensions that describe Russian influence over a group or individual. The different dimensions of this typology are not always easily measurable (and we do not attempt to systematically measure them for particular individuals or groups), but they do offer a descriptive language to consider the range and forms of Russian influence, and Russia's ability to achieve its objectives through a proxy or allied group.

The first dimension is Russia's *control* over an actor, meaning the extent to which the actor will follow the orders or expressed desires of the Russian government. At the more effective end, Putin's administration clearly has higher degrees of control over parts of the Russian state, or of proxies that are fully funded by Russia and staffed by former Russian officials. Russia might have more moderate control over groups to which it provides significant funding or that it can otherwise control or blackmail. There are also many organizations that Russia might fund but over which it has little control, such as the Front National in France.

The second dimension is the *alignment of interests* between Russia and a given group or individual. If an individual or organization shares an interest in a particular policy, there is less need for Russia to exercise control on that particular issue. For example, one analyst in Latvia highlighted that Russia had decided to support anti–gay marriage legislation in order to gain support from conservative parties,[41] and Russia allegedly funded anti-fracking groups in Bulgaria to prevent increases in local energy production and maintain demand for Russian gas.[42] When groups that share an interest with Russia provide at least a tacti-

[41] Phone call with Latvian analyst, December 2015.

[42] Kerin Hope, "Bulgarians See Russian Hand in Anti-Shale Protests," *Financial Times*, November 30, 2014.

cal benefit, it is sufficient for Russia to provide resources to that group to pursue Russian objectives.

The third dimension is *leverage*—the ability of the individual or group to actually influence policy in Russia's favor. Lower-leverage actors, such as Russian-funded think tanks in Western countries, have only minimal influence on policy. Higher-leverage actors, such as the Front National in France or Viktor Orban in Hungary, might have greater ability to influence policy in their host nations. Given that countries in Europe are democracies with political checks and balances and strong rules of law, institutional and structural factors limit the ability of even the most powerful individuals or groups to change policy.

One finding of this report is that there is typically a trade-off between the leverage of a group and Russia's alignment of interests or degree of control over that group. Groups with direct links to Russia's intelligence apparatus can be easily controlled but tend not to have particularly high leverage. Conversely, Russia's control over more-powerful actors, such as the Front National or Viktor Orban, is far more limited, as is its alignment of interests. Efforts by Russia to increase the leverage of a proxy can go along with diminishing control or alignment of interests, and efforts to increase Russia's control might diminish the leverage of a given proxy.

A final consideration is that Russia faces significant principal-agent challenges. Many of its proxies might have other agendas, alternative interests, and an ability to hide their activities from Russian officials. One telling example is the separatists in Ukraine. Igor Girkin (also known as Strelkov), a prominent separatist fighter who claimed to be a former Federal Security Service (FSB) agent, was apparently removed from his position and returned to Moscow, where he has become a prominent critic of Russia's policy in Ukraine.[43] Russia cannot always control individuals and groups, and even when interests are generally aligned, Russia will exert significant pressure.

---

[43] Andrew Roth, "Former Russian Rebels Trade War in Ukraine for Posh Life in Moscow," *Washington Post*, September 16, 2015.

Conclusion

Russian hostile measures are of increasing concern to policymakers, especially following Russian influence in recent U.S. activities. Russia has various interests that lead it to pursue hostile measures in Europe, including its goals of protecting the regime, maintaining its sphere of influence, and pursuing great-power status. It appears to adopt a soft strategy, in which several different hostile measures are used simultaneously with broad foreign policy goals in mind but without a specific causal logic of success. Defending against hostile measures requires looking at Russia's tactics in specific regions, which is the focus of the remainder of this report.

CHAPTER TWO

Russian Hostile Measures in the Baltics

The Baltic states are commonly said to be particularly vulnerable to Russian hostile measures. They are the only countries in the EU and NATO that were formerly republics of the Soviet Union. Estonia, Latvia, and Lithuania all share a border with Russia, and Estonia and Latvia also have significant numbers of Russian speakers, including ethnic Russians, Ukrainians, Belarusians, and others, many of whom migrated there during the Soviet period. Although many of these people have integrated into these countries, some remain loyal to Russia or sympathetic to Russian views. Russia retains significant economic ties with these regions, including ownership of transportation businesses, sale of oil and gas, and organized crime with reported links to the Russian state.

Influence in the Baltics could serve a range of Russian interests, including undermining the EU and NATO and limiting the perceived threat of NATO and EU enlargement. Russian officials and analysts downplay Russia's interests in the Baltics, although Russian influence activities do remain, as will be discussed. The rhetoric about protecting Russian speakers might also be largely instrumental, a justification for Russian influence rather than a goal on its own.

To understand what type of NATO response is necessary, it is first critical to understand what hostile measures Russia could impose. Given the ongoing threat of Russian hostile measures, enhanced support from U.S. and allied nonmilitary government agencies, along with a carefully scoped conventional military presence, could be valuable.

Motives

According to some, the five postulated strategic objectives described in Chapter One suggest a wide variety of motives for Russia to interfere in the Baltic states. Other factors mitigate Russia's desire to gain influence in the Baltic states, especially compared with other former Soviet republics. Nevertheless, Russia does appear motivated to seek some degree of influence, which could undermine stability and prosperity in the Baltics. Here, we discuss five possible motives for Russia to take hostile measures.

First, the Baltic states could be the most vulnerable point of the EU and NATO. Russia might be able to demonstrate the failure of EU and NATO guarantees in the Baltic states far more readily than elsewhere in the alliance, thereby undermining the alliance's credibility. From this perspective, the Baltic countries are of no intrinsically greater interest than other EU or NATO members, but the vulnerabilities related to geography, presence of Russian speakers, economic ties, and other factors (discussed in the "Opportunities" section of this chapter), might make it less costly and risky for Russia to undermine the Baltics than to undermine other EU or NATO members. In particular, Russia might have reason to believe that NATO is more likely to abandon the Baltics than other members. In discussions, U.S. policy analysts sometimes questioned the original admission of the Baltics to NATO while acknowledging that they should be defended now as members of the alliance. President Barack Obama emphasized NATO's commitment to the Baltic states in his speech in Talinn in September 2014, noting, "We'll be here for Estonia. We will be here for Latvia. We will be here for Lithuania. You lost your independence once before. With NATO, you will never lose it again."[1] Even with this reassurance, there is some degree of doubt as to whether NATO will truly follow up on its commitment. Russia also might be able to manipulate the local Russian-speaking population in ways that make it appear as if Russian actions represent local rebellion rather than outside aggression. NATO

[1] White House, "Remarks by President Obama to the People of Estonia," September 3, 2014.

member states might decide not to respond to Russian aggression in the region if it is perceived as a purely internal matter, or they might use Russia's denial of involvement in such conflicts as an excuse to avoid becoming involved.

Second, Russia might perceive a threat from the Baltic states and use hostile measures to counter this threat. It is possible that Russian military and political leaders could view the Baltic countries as a launching ground for color revolutions, democracy promotion, or other activities that Russia perceives as threatening to the regime. Russian leaders have not highlighted this threat from the Baltics thus far, but this general line of thinking is consistent with their discourse. For example, Russia's chief of the General Staff noted that the NATO alliance has moved much closer to Russian borders through the integration of the Baltic countries and that NATO has proposed deploying forces into the Baltic states and other countries on its eastern flank.[2]

Third, the use of hostile measures in the Baltics could offer the regime a possible political boost. The Maidan protests in Ukraine and Russia's annexation of Crimea coincided with a significant increase in approval for Putin, and Putin's government might similarly be able to draw on public support during political crisis in the Baltics.[3] Given the Baltic states' large numbers of Russian speakers and proximity to Russia, Moscow could draw on a number of messages to justify its support for Russian speakers and boost public opinion for the regime. However, the regime does not entirely control Russian public opinion, and, as already discussed, the regime appears concerned about public uprising. Even though Russians widely supported the annexation of Crimea, direct Rus-

---

[2] Gorenberg, 2015.

[3] Olga Oliker, Christopher Chivvis, et al. note,

> The Russian public's continuing support of its government in the foreign policy sphere is thus likely a result of both a general predilection to support the government on such issues and the appeal of the specific messages that the Kremlin has used. Nationalism and rebirth are appealing concepts.

Olga Oliker, Christopher S. Chivvis, Keith Crane, Olesya Tkacheva, and Scott Boston, *Russian Foreign Policy in Historical and Current Context*, Santa Monica, Calif.: RAND Corporation, PE-144-A, 2015, p. 20. See also Peter Pomerantsev, *Nothing Is True and Everything Is Possible: The Surreal Heart of the New Russia*, New York: Public Affairs, 2015.

sian military involvement in Ukraine was unpopular in Russia, perhaps partly because of the longstanding rhetoric of brotherhood between the two.[4] Therefore, war or other forms of Russian aggression in the Baltics might not be popular and could lead to a significant backlash in public opinion, depending on the political context.

A fourth potential reason Russia might undertake hostile measures in the Baltics would be to exert influence on their domestic and foreign policies because of their status as former Soviet republics on Russia's border. Although some intrinsic Russian desire for influence in the Baltics likely does exist, as discussed in Chapter One, it is also worth noting that such desires are likely significantly higher for Ukraine, Moldova, and other former Soviet states.[5] Robert Person, for example, observes,

> Ukraine is special for Russia. The Baltics are a different story. . . . Russia and Russians have long recognized that the Baltics are culturally and historically distinct from Russia, according to surveys and interviews I've conducted across Russia and Latvia.[6]

Indeed, Russian speakers who emigrated to the Baltics interacted with the native societies in different ways than occurred in other Soviet republics, especially in terms of seeking greater integration into local society.[7] Although Russia did initially oppose the accession of the Baltic states to NATO, its response to these countries' NATO enlargement in 2004 was rather muted.[8] Russian analysts we interviewed highlighted that the Baltic countries were no longer in Russia's sphere of influence,

---

4 Harley Balzer, "The Ukraine Invasion and Public Opinion," *Georgetown Journal of International Affairs*, March 20, 2015.

5 We thank Stephanie Pezard, Katya Migacheva, and Brenna Allen for unpublished research related to this report.

6 Robert Person, "6 Reasons Not to Worry About Russia Invading the Baltics," *Washington Post*, November 12, 2015.

7 David Laitin, *Identity in Formation*, Ithaca, N.Y.: Cornell University Press, 1998.

8 See Mankoff, 2012, p. 160; Vladimir Putin, "Annual Address to the Federal Assembly of the Russian Federation," Moscow, April 25, 2005.

and some U.S. analysts of Russia made similar claims.[9] Russian media has also downplayed the importance of the region in the context of NATO's growing presence there.[10] Still, such Russian claims do not mean that there is no threat of Russian hostile measures related to a desire for influence. Such claims might instead mean that these factors could mitigate Russia's hostile measures in the region or that hostile measures might be developed based on other interests.

A final potential motive is Russia's interest in supporting or influencing ethnic Russian and Russian-speaking populations. Russia's foreign policy articulates a desire to protect its "compatriots" abroad, meaning Russian speakers and former Soviet citizens, including by "preserving and promoting the Russian language and culture," enabling "compatriots to better realize their rights in their countries of residence," and facilitating "the preservation of the Russian diaspora's identity and its ties with the historical homeland."[11] Outreach to compatriots is one explicit purpose of the Federal Russian *Rossotrudnichestvo* agency, which is active in the Baltics.[12] There is a question, however, of whether reaching out to these compatriots is an end in itself, as opposed to a means to achieve other objectives. For example, Russian scholar Zevelev explains,

> Moscow has always treated the protection of rights and interests of Russians and Russian-speaking minorities much more as an instru-

[9] Discussions with U.S. and Russian think tank analysts, Washington D.C., April 2016.

[10] A Russian media response to a RAND report on the possible invasion of the Baltics noted,

> All of the announcements of the White House about a possible Russian attack on the Baltic States are political and military nonsense. Russia does not intend to return the unfortunate producers of sprats [a sardine-like fish popular in the Baltics] to the fold. There is no need for them.

Vladimir Ivanov, "Washington's Baltic Role: America Strengthens Eastern Flank of NATO," *The Independent* (in Russian), March 4, 2016.

[11] Russian Federation, Ministry of Foreign Affairs, Foreign Policy Concept of the Russian Federation, November 30, 2016, clause 45.

[12] Federal Agency for the Commonwealth of Independent States, Compatriots Living Abroad, and International Humanitarian Cooperation, "About *Rossotrudnichestvo*," webpage, undated.

> ment of securing leadership in the territory of the former Soviet Union rather than as a goal in itself. . . . Moscow generally believes that it should not drop the problem of Russian nationals abroad from the foreign policy agenda, but it has never prioritized this issue. Relations with Latvia and Estonia are an exception to this rule, but here, too, in moments of crisis Russia's economic interests compel it to confine its actions to loud rhetoric, as was the case in the conflict with Tallinn over the Bronze Soldier monument.[13]

A report from the Baltic states similarly notes, "Russia wants to use compatriots living abroad as a geopolitical entity that defends Russia's interests, regardless of the compatriots' home countries or other identities. In this case, Russia's strategy lies in an even deeper-rooted tradition of Russian foreign policy, one that is percepted [sic] as the notion of (post)imperial control."[14] Thus, although Russia's policy in the Baltic states did make Russian speakers a priority, it remains unclear whether and to what extent this policy was a means to other ends or reflected a real underlying foreign policy goal.[15] Either way, Russia's desire to influence the Russian speakers remains a matter of concern for the stability of the Baltic states.

Opportunities

Several factors make the Baltic countries especially vulnerable to Russian hostile measures. The main factors are their geography and history within the Soviet Union. The Baltics were independent countries prior to World War II, occupied by the Soviet Union in 1939–1940, captured by Nazi Germany in 1941, and retaken by the Soviet Union in 1944.

---

[13] Igor Zevelev, "Russia's Policy Towards Compatriots in the Former Soviet Union," *Russia in Global Affairs*, Vol. 6, No. 1, January–March 2008, p. 55.

[14] Pelnens, 2009, p. 21.

[15] Even in Ukraine, although Putin justified Russia's actions as based on concern for the Russian population, this set of claims appears to have been made only in retrospect and does not offer a reliable explanation of Russia's behavior. See Daniel Treisman, "Why Putin Took Crimea: The Gambler in the Kremlin," *Foreign Affairs*, May–June 2016.

They were made Soviet republics, and although they continued to resist Soviet rule, a number of close links were created between the Baltic states and Russia that set the context for a number of issues we will describe later. In 1991, the three Baltic countries were among the first republics to declare independence from the Soviet Union, after which all three adopted policies of legal continuity with the pre-Soviet republics. The majority, or "titular," ethnic group sought to consolidate control over the identity of the country and achieve integration with Western institutions, partly out of fear of a renewed threat from Russia.[16]

Beyond the Russian influence originating from the shared Soviet history, there is the fact that Estonia, Latvia, and Lithuania all share a border with Russia, which gives Russia a direct means of exerting pressure over these countries. In Estonia, for example, Eston Kohver, an Estonian internal security service official, was seized by the FSB and accused of espionage in September 2014. Estonian authorities insisted that he was on the Estonian side of the border and declared that he was captured because he was investigating organized crime involving FSB agents. Kohver was eventually freed in an exchange, but these events highlight how Russia could use a shared border to put pressure on the significantly smaller Baltic states.[17] Lithuania has a somewhat different geography, bordering Belarus and the Russian enclave of Kaliningrad. Therefore, Russian military and civilian vehicles must transit Lithuania, and Russian citizens in Kaliningrad regularly transit Lithuania and Poland. Russia and Lithuania have longstanding agreements in place for military transit of Russian troops through Lithuania, which do not appear to be diminishing.[18] There are also agreements in place for Russians to travel to Kaliningrad without a Schengen visa (valid for travel elsewhere in the EU) and for Russians living in Kaliningrad to travel

---

16 Andres Kasekamp, *A History of the Baltics*, New York: Palgrave Macmillan, 2010.

17 Discussions with Estonian officials, Tallinn, July 2015; discussions with think tank analysts, London, February 2016; "Russia and Estonia 'Exchange Spies' After Kohver Row," BBC, September 26, 2015.

18 Russia is apparently constrained to maintain military transports in Lithuania because Russian conscripts are not permitted to have travel documents. RAND MSW symposium, Cambridge, United Kingdom, February 3, 2016.

visa-free into northeastern Poland.[19] Although the imperative for Russia to transit Lithuania to access Kaliningrad does give Lithuania additional leverage over Russia, it also might increase the risk that Russia could take aggressive action if Lithuania cut off access to Kaliningrad.

Russian-Speaking Minorities

The substantial number of Russian speakers in Estonia and Latvia offers perhaps the greatest opportunity for Russian hostile measures. Many ethnic Russians, as well as Soviet citizens of other nationalities, migrated to Estonia and Latvia during the Soviet period. They and their descendants are often referred to as "Russian speakers," and they have, to some degree, become a cohesive ethnic and political group.[20] The proportion of Russian speakers has declined since 1989,[21] but this group still represents a sizable percentage of the population. In 2011, 30 percent of the Estonian population identified Russian as their mother tongue, and 25 percent of the population identified as ethnically Russian.[22] Approximately 27 percent of the Latvian population identified as ethnically Russian and, based on national origins, approximately 35 percent of the population are Russian speakers.[23] The Russian and Russian-speaking population in Lithuania is much smaller—ethnic Russians made up only 5.8 percent of the population in 2011.[24]

---

[19] Ingmar Oldberg, *Kaliningrad's Difficult Plight between Moscow and Europe*, Stockholm: Swedish Institute of International Affairs, No. 2, 2015.

[20] Laitin, 1998.

[21] In 1989, Estonia was 62 percent ethnic Estonian and Latvia was 52 percent ethnic Latvian; in 2011, these numbers were 70 percent and 62 percent, respectively. Kasekamp, 2010, p. 155.

[22] Statistical Office of Estonia, Central Statistical Bureau of Latvia, and Statistics Lithuania, *2011 Population and Housing Censuses in Estonia, Latvia, and Lithuania*, 2015, pp. 10, 24.

[23] The total number of Russian speakers includes Russians, Belarusians, Ukrainians, and Poles relative to the overall population of Latvia. See Statistical Office of Estonia, Central Statistical Bureau of Latvia, and Statistics Lithuania, 2015, p. 24. Grigas observes 34 percent Russian speakers in Latvia. Agnia Grigas, "The New Generation of Baltic Russian Speakers," EurActiv.com, November 28, 2014.

[24] Statistical Office of Estonia, Central Statistical Bureau of Latvia, and Statistics Lithuania, 2015, p. 24.

The Russian speakers in Estonia and Latvia are concentrated in the capital and eastern portions of both countries (see Figure 2.1).

Russian speakers have been only partially integrated into Estonia and Latvia, and many are frustrated with their position in these societ-

Figure 2.1
Concentrations of Russian Speakers in the Baltics

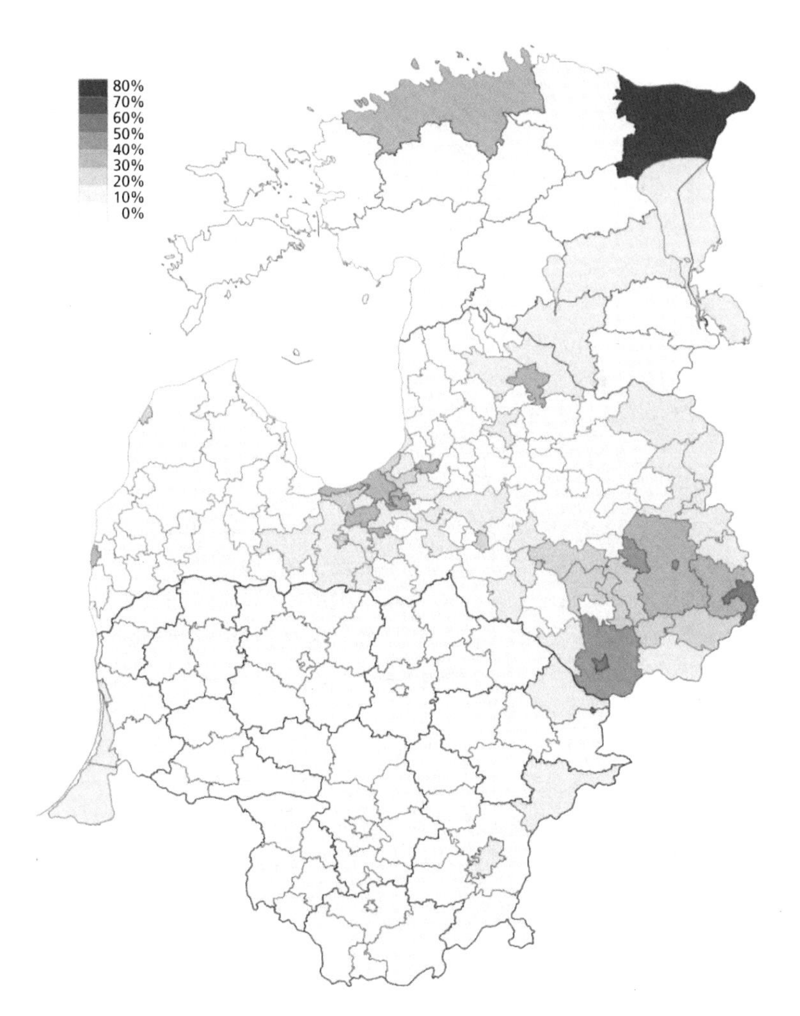

SOURCE: Xil, "Russians in Baltic States, 2011," licensed under CC BY-SA 3.0 via Commons, uploaded February 19, 2015.
RAND *RR1793-2.1*

ies, especially regarding citizenship, respect for Russian speakers as a minority, and income inequality. The decision by Estonia and Latvia to maintain legal continuity meant that only individuals who were the descendants of citizens of the pre–World War II republics were automatically made citizens of the new republics. At the same time as Estonia and Latvia emphasized their adherence to liberal principles, in part because of their desire for integration into the EU and NATO, there was significant nationalist rhetoric stating that Russian speakers were not welcome, encouraging voluntary repatriation to Russia, and making it difficult for Russian speakers to become Estonian and Latvian citizens.[25] Although Estonia and Latvia subsequently liberalized their citizenship laws—again, partly because of the process of integration into the EU—many Russian speakers have not yet become citizens of their home country. In 2011, approximately 54 percent of ethnic Russians in Estonia were Estonian citizens, 24 percent held Russian citizenship, and 21 percent had "undetermined" citizenship, which generally means they were issued "alien" papers that permit work and visa-free travel throughout the EU.[26] In Latvia in 2011, 14.3 percent of the total population did not have citizenship, and only 1.63 percent of the total population were citizens of Russia.[27] Beyond the legal aspects of citizenship, Russian speakers have a sense—perhaps greater in Estonia than in the other two Baltic states—that they are mistreated. In a 2009 survey, for example, 17 percent of Russians (not distinguished between ethnic Russians or Russian speakers) in Estonia stated that they had been discriminated against, compared with 4 percent and

---

[25] Laitin writes,

> The leaders of the restored Estonian and Latvian republics were unequivocal nationalizers. They were willing to accord internationally recognized human rights to nontitulars, but they had no intention of allowing them to play key roles in the rebuilding of the cultural foundation of the restored nation-states.

Laitin, 1998, pp. 93, 167.

[26] See the results of the 2011 Census in Estonia. Statistics Estonia, "Statistical Database," question PC0442, undated.

[27] Central Statistics Bureau of Latvia, "ISG09. Population of Latvia by Citizenship at the Beginning of the Year," undated-a.

5 percent in Lithuania and Latvia, respectively. In this survey, Russians were especially concerned about discrimination in looking for work.[28]

Indeed, Russian speakers in Estonia and Latvia do appear to face poor economic prospects, especially in rural areas. A 2011 paper noted lower incomes for Russian speakers in Estonia and observed that learning the titular language (i.e., Estonian or Latvian) did not increase the economic prospects of Russian speakers in Estonia or Latvia despite similar levels of education and labor participation. Instead, learning English was a better predictor of increased wages, implying that Russian speakers faced persistent discrimination that they could do little to combat.[29] Rural areas dominated by the Russian minority face especially serious economic challenges. For example, in 2011, unemployment in Estonia was 5.3 percent overall, compared with 8.3 percent in the mainly Russian-speaking Ida-Viru county in the far northeast of the country. In Latvia, unemployment in the country has improved from 16.5 percent in 2011 to 10.1 percent in 2015 but remains at 19 percent in the predominately Russian-speaking region of Latgale.[30] However, incomes in rural regions of Estonia and Latvia are still higher than those of neighboring regions of Russia.[31]

---

[28] European Union Agency for Fundamental Rights, *EU-MIDIS: European Union Minorities and Discrimination Survey*, Luxembourg: Publications Office of the European Union, December 2009, p. 36.

[29] The paper explained that "the Russian minority still suffers from the glass-ceiling effect," and noted that since 1990 in Estonia there was an "unexplained income differential, about 10–15 percent in favor of ethnic Estonians." Ott Toomet, "Learn English, Not the Local Language! Ethnic Russians in the Baltic States," *American Economic Review*, Vol. 101, No. 3, 2011, pp. 531, 529.

[30] Central Statistics Bureau of Latvia, "NBG04. Activity Rate, Employment Rate and Unemployment Rate by Statistical Region," undated-b.

[31] The gross domestic product (GDP) per capita of the key Russian majority regions of Estonia and Latvia, Ida-Viru and Latgale, are also higher than the neighboring region of Leningrad and Pskov in Russia ($12,975 for Ida Viru and $11,886 for Leningrad; $7,866 for Latgale, and $5,227 for Pskov Region in 2013. See Statistics Estonia, undated; Central Statistics Bureau of Latvia, "Statistics Database," undated-c; Knoema, "Leningrad Region—Gross Regional Product per Capita," undated-a; Knoema, "Pskov Region—Gross Regional Product per Capita," undated-b. Exchange rates from Internal Revenue Service, "Yearly Average Currency Exchange Rates," updated January 15, 2016. For inequality data, see World Bank, "GINI Index (World Bank Estimate)," undated-a.

Russian speakers in the Baltics do not necessarily comprise a reliable fifth column for Russia, however. Some Russian speakers might share Russian interests, indicating some alignment, but there are few indications of control or leverage from Moscow. Levels of assimilation vary among Russian speakers, but many are well integrated into their home country. In Estonia, one study claimed that 37 percent of Russian speakers are either "successfully integrated" or are "Russian-speaking patriot[s]" while many of those who have not integrated are older, former Soviet citizens.[32] It is generally agreed that integration has gone better in Latvia—one-third of marriages are interethnic, 15 percent of those serving in Latvia's armed forces are Russian speakers, and 64 percent of Russian speakers identified as either strongly or moderately "affiliated to Latvia" in a 2014 survey.[33] In interviews with the media, Russian speakers in border regions have consistently downplayed the potential for separatism or rebellion. For example, in Estonia, the former leader of the 1993 Narva autonomy referendum rapidly accepted "the normality of using Estonian in the offices" and appeared relatively content in a recent interview with the political status quo.[34] According to Radio Free Europe/Radio

[32] Juhan Kivirähk, *Integrating Estonia's Russian-Speaking Population: Findings of National Defense Opinion Surveys*, International Centre for Defence and Security, December 2014, pp. 8–9. Laitin also describes nonlinguistic forms of assimilation, in which Russian speakers in Estonia behave more like Estonians than Russians. Laitin, 1998. See also Aksel Kirch, Marika Kirch, and Tarmo Tuisk, "Russians in the Baltic States: To Be or Not to Be?" *Journal of Baltic Studies*, Vol. 24, No. 2, Summer 1993.

[33] Interviews with former Latvian government official and analyst, Riga, July 2015; "Two-Thirds 'Loyal' to Latvia in Minority Poll," Latvian Public Broadcasting English-Language Service, August 26, 2014.

[34] Laitin 1999, p. 153. A *Wall Street Journal* article in July 2014 described the leader of the referendum:

> I caught up with [Mr. Chuikin] in Tallinn. Over the past two decades, he had gone into business and retired. Now in his early 60s, he still hasn't learned Estonian or obtained citizenship. He shows off his Russian passport, even though he has lived most of his life in Estonia. He's monolingual and gripes that the young Estonian hostess at the coffee shop where we meet doesn't speak Russian. Mr. Chuikin recently moved to a new suburban home in Tallinn next door to his daughter, who married an ethnic Estonian and has bilingual children. His other daughter moved to Stockholm, married a Swede and

Liberty (RFE/RL), a Russian-speaking resident of Narva who traveled across the border to Russia noted, "When you cross into Ivangorod, straight away you can see the atmosphere there[.] Who is going to want to join that?"[35]

Russian speakers in the Baltics do largely consume media originating within Russia. Estonian analysts note, "In Estonia, Estonians and non-Estonians live in different information spaces, often with contrasting content . . . Most of the Russian-speaking population derives its information and views on history and current events from Russian television channels that are directly subordinate to the Kremlin and can be used as a mechanism of propaganda."[36] Similarly, Andis Kudors, a Latvian analyst, observes,

> Russia, while implementing its public diplomacy in Latvia, uses a selective approach—i.e., official Moscow addresses mainly target the Russian audience in Latvia, ignoring Latvians. Considering the presence of numerous Russian media and their popularity with the Russian audience in the Baltic states, such an approach promotes maintaining the split in society, and hampers the society integration process.[37]

While many Russian speakers clearly trust and prefer media originating in Russia, many do not blindly accept Russian media messages.

> is raising her own multilingual kids. Every few weeks, he makes the quick flight over the Baltic Sea to see them. His Estonian residency permit lets him travel around Europe without a visa. He enjoys a European life and admires Vladimir Putin.

Matthew Kaminski, "The Town Where the Russian Dilemma Lives," *Wall Street Journal*, July 4, 2014.

[35] See Tom Balmforth, "Russians of Narva Not Seeking 'Liberation' by Moscow," Radio Free Europe/Radio Liberty, April 4, 2014.

[36] Mike Winnerstig, *Tools of Destabilization: Russian Soft Power and Non-Military in the Baltic States*, Swedish Defence Research Agency, December 2014. pp. 52–53.

[37] Andis Kudors, "Reinventing View to the Russian Media and Compatriot Policy in the Baltic States," in Artis Pabriks and Andis Kudors, eds., *The War in Ukraine: Lessons for Europe*, Riga, Latvia: Centre for Easter European Policy Studies, University of Latvia Press, 2015, p. 163.

In one survey, 43 percent of the minority respondents in Latvia said Russian-language media is objective while 36 percent disagreed.[38]

There is also activity on social media in the Baltic states that might serve Russia's interests, but campaigns that appear friendly to Russia have not been attributed to the Russian state. For example, in Latvia, a popular website that had been disseminating cat videos and other benign content shifted to an "everything is bad in Latvia" format. The content appeared to have the same general approach that Russia might use. However, a Latvian tech blogger, Janis Polis, tracked the campaign back to a Russian-speaking Latvian member of the European Parliament.[39] Similarly, Estonian officials attributed many negative social media campaigns to local Russian speakers.[40] The absence of a significant, clear, or concerted campaign initiated by Moscow does not mean that one could not emerge in the future, however.

Furthermore, a range of Russian-speaking organizations in Estonia and Latvia pursue increased language rights, citizenship, and other issues of concern to the Russian-speaking community.[41] Perhaps the most prominent of these are political parties. In Estonia, the Centre Party receives most of its support from Russian speakers and was the second-largest party in the 2015 elections. After longtime spearhead Edgar Savisaar left the party leadership in November 2016, his replacement, Jüri Ratas, joined the governing coalition and became Prime Minister. There appears to have been no significant change in Estonia's

---

[38] "Two Thirds 'Loyal' to Latvia in Minority Poll," 2014.

[39] Interview with Latvian social media researcher, Riga, January 2017; "Mystery Website Producer Has Ties to Harmony, LTV Reports" Latvian Public Broadcasting English-Language Service, January 9, 2017.

[40] Todd Helmus, Elizabeth Bodine-Baron, Andrew Radin, Madeline Magnuson, Joshua Mendelsohn, Bill Marcellino, Andriy Bega, and Zev Winkelman, *Russia Social Media Influence: Understanding Russian Propaganda in Eastern Europe*, Santa Monica, Calif.: RAND Corporation, RR-2237-OSD, 2018, p. 50.

[41] For example, in Estonia there is the "Russian School of Estonia" that advocates Russian-language education, and the Estonian Aliens Union, which seeks changes on citizenship issues. See Русская Школа Эстонии [Russian School of Estonia], homepage, undated; Union of Stateless People of Estonia, homepage, undated.

foreign and security policy associated with Ratas taking power.[42] The Centre Party formerly had an agreement with Putin's party, United Russia, which the Centre Party has declined to activate but also did not repudiate, likely in part to remain attractive to Russian-speaking voters.[43] In Latvia, popular Riga mayor Nils Usakov led Harmony Centre, which is the largest political party in the country, receiving 28 percent of the vote in the 2014 elections.[44] Like the Centre Party, Harmony Centre is said to have formal ties with United Russia but appears reluctant to admit them to the coalition for fear it would be influenced by Russia.[45] In 2007, the President of Latvia claimed to have "confidential information on [Harmony Centre's] financial resources that causes concern about the party's loyalty to the interests of the state of Latvia," but declined to elaborate, noting classified sources.[46] One analyst in Latvia emphasized that reluctance to cut ties with United Russia was causing Harmony Centre to lose votes among ethnic Latvians who might otherwise be willing to support it.[47]

Beyond the political parties, many pro-Russian NGOs advocate on behalf of the interests of Russian speakers, and some of these might have ties with Russia intelligence or security institutions. The

---

[42] According to the Legal Information Centre for Human Rights, the Centre Party in 2011 was "supported by 81% of Russian-speaking citizens and 11% of enfranchised Estonians," who are generally lower-income and older Estonians. Vadim Poleshchuk, *Russian-Speaking Population of Estonia in 2014: Monitoring Report*, Tallinn, Estonia: Legal Information Centre for Human Rights, 2014, pp. 14, 17; Richard Martyn-Hemphill, "Estonia's New Premier Comes from Party with Links to Russia," *New York Times*, November 20, 2016; discussions with Estonian officials, Washington and Talinn, November 2015 and January 2017, and December 2017.

[43] "Ratas: Center Party Not Planning to Give Up Protocol with United Russia," ERR.ee, October 9, 2017.

[44] Licia Cianetti, "The Governing Parties Survived Latvia's Election, but the Issue of the Country's Russian-Speaking Minority Remains Centre-Stage," London School of Economics blog, October 10, 2014.

[45] Richard Milne, "Party with Ties to Putin Pushes Ahead in Estonian Polls," *Financial Times*, February 27, 2015; "How to Deal with Harmony," *The Economist*, October 5, 2014.

[46] Winnerstig, 2014, p. 85.

[47] Discussion with Latvian think tank analyst, Riga, July 2015.

Baltic states' security services publish annual reports warning of Russian influence through these organizations, although, in practice, the popularity, legitimacy, and level of Russian control likely varies. More-radical groups—such as "anti-fascist" organizations under the umbrella group *Mir Bez Natsizma* or World Without Nazism—appear to have limited support. Other organizations have more credibility, such as the Legal Information Centre for Human Rights in Estonia, which partners with Amnesty International.[48] Many groups are focused on creating or disseminating an interpretation of World War II more in line with the Soviet and Russian version of history, or on promoting adoption of Russian as an official language.[49] According to the Estonian Internal Security Service (KAPO), Russian-speaking Estonians attend Russian government–sponsored youth camps emphasizing Russia's victory in World War II, and are reported to have links with Russian military intelligence.[50]

Estonian and Latvian officials claim that Russia could organize a protest or opposition if it so desired while downplaying the general receptivity of "our Russians" to provocation by Russia.[51] Although Baltic officials tend not to recognize the grievances of Russian speakers, perhaps because of nationalist narratives, it is important to recognize that support for separatism or similar activities by Russian speakers appears limited. A 2015 *Foreign Affairs* article notes that in the mainly Russian city of Narva,

> Pro-Russian protests are few and far between. When they do occur, they tend to garner 'about 20 to 40 people,' and sometimes a third to a half of those are from the media and/or Estonian law

---

48 Pelnens, 2009, pp. 72–73.

49 Inga Spriņģe, Donata Motuzaite, Gunita Gailāne, "Money from Russia: Spreading Democracy in Latvia, Kremlin Style," *Re: Baltica*, March 19, 2012.

50 KAPO, "Annual Review," 2013, p. 11.

51 Discussions with Baltic officials, Tallinn, Estonia; Riga, Latvia; Cambridge, United Kingdom; and Washington D.C., July and November 2015, February 2016.

enforcement agencies, according to Arnold Sinisalu, the Director General of KAPO.[52]

Agnia Grigas notes a few examples of individuals supporting separatism in Latvia but writes that "separatist sentiments are generally the exception rather than norm among the Russian minority, both in Latgale and in Latvia as a whole."[53] Although Russia might be able to encourage small numbers of Estonians or Latvians to adopt a policy of separatism or other major destabilization, there is limited evidence of large-scale, spontaneous local support for such movements.

The confidence of Estonian and Latvian officials also reflects the relatively well-developed security forces of these countries. Their focus and ability to monitor Russian groups is demonstrated by the publications from the internal security services in these countries and reflected in discussions with officials and analysts.[54] In some cases, this attention might become counterproductive when groups that reflect the genuine and legitimate concerns of the Russian-speaking community are identified as tools of Russia—Estonia's Centre Party and Latvia's Harmony Centre might be the best examples. We discuss the capabilities of the Baltic security forces in assessing the threat of covert violent action, but it is difficult to know the extent to which different Baltic countries can monitor the activities of pro-Russian groups. (This would be a valuable area for further study.)

---

[52] Michael Weiss, "The Estonian Spymasters," *Foreign Affairs*, June 3, 2014.

[53] Grigas offers two examples of separatism being discussed:

> In April 2014, a sparsely attended rally took place in front of the Latvian embassy in Moscow calling for Latgale to become part of Russia and in early 2015 a Latvian activist was arrested by the authorities because he was collecting signatures for a petition sponsored by the website avaaz.org for Latvia to be annexed by Russia.

Agnia Grigas, *Beyond Crimea: The New Russian Empire*, New Haven, Conn.: Yale University Press, 2016, p. 167.

[54] See KAPO, 2013; Latvia Security Police, "Annual Report," 2013; discussions with Baltic and U.S. officials and analysts, Tallinn; Riga; and Cambridge, United Kingdom, July 2015 and February 2016.

Other Minority Groups

Other minority groups in the Baltic states could be vulnerable to Russian assistance. In Lithuania, ethnic Russians make up less than 6 percent of the population, and ethnic Poles are approximately 6.6 percent.[55] Many ethnic Poles speak Russian, and there have been some efforts to combine the political forces of these groups. For example, Waldemar Tomaszewski, leader of the Electoral Action of Poles in Lithuania, has also sought to court the ethnic Russian vote.[56] Baltics officials and analysts downplay the risk of Russian mobilization of these groups, although this could change in the future.[57]

Energy Imports

Another potential point of vulnerability is the Baltic states' dependence on Russian energy, although this dependence appears to have substantially diminished in recent years, especially in Lithuania.[58]

Prior to the opening of the Klaipeda liquefied natural gas (LNG) terminal in 2015, the Baltic states were almost exclusively dependent on Russia for natural gas.[59] Natural gas represents more than 30 percent of non-oil energy usage in Latvia and Lithuania, making substitution with other fuels somewhat difficult. "Protected" customers, such as residential users of natural gas, represent more than 70 percent of consumption in Estonia and Latvia and approximately 20 percent in Lithuania, meaning that a shortage of gas could have serious

---

[55] Statistical Office of Estonia, Central Statistical Bureau of Latvia, and Statistics Lithuania, 2015, p. 24.

[56] "Stirring the Pot," *The Economist*, May 3, 2015.

[57] Discussions with analysts and officials, Cambridge, United Kingdom, February 2016.

[58] One Lithuanian official emphasized that the threat of energy dependence to Lithuania had significantly diminished because of construction of Klaipeda and alternative sources of electricity, meaning that the major threat to Lithuania's security lay with a more conventional military attack. RAND MSW symposium, Cambridge, United Kingdom, February 3, 2016.

[59] See European Commission, "Communication from the Commission to the European Parliament and the Council on the Short-Term Resilience of the European Gas System Preparedness for a Possible Disruption of Supplies from the East During the Fall and Winter of 2014/2015," SWD(2014) 322 final, Brussels, October 16, 2014.

repercussions for those populations.[60] In response to concern about a cutoff of Russian gas, Lithuania developed a major policy of achieving energy independence from Russia, including such elements as the construction of the Klaipeda terminal and the establishment of the NATO Energy Security Center of Excellence.[61] At full production, the Klaipeda terminal has the capacity to supply 4 billion cubic meters of gas, which Lithuanian officials claim will be able to supply up to 90 percent of requirements in the other Baltic states.[62] Latvia also has substantial natural gas storage facilities, which would help all three countries weather a short-term cutoff in supply.[63]

The Baltic states continue to import oil and electricity from Russia. Alternative imports for crude and refined oil products, however, are far less constrained by the existing infrastructure than natural gas imports. Barring a cutoff in access to the Baltic Sea in the case of war, the Baltic states could easily find alternative crude and refined oil suppliers if Russia were to halt imports.[64] The Baltic countries also share their electrical grids with Belarus, Kaliningrad, and northwestern Russia, and there remain significant imports of electricity from Russia. Lithuania, for example, received 23 percent of its electricity from Russia in 2012; Latvia received 17 percent, some of which was resold to Estonia.[65]

A Russian cutoff of electricity to the Baltic states could cause disruption to the three nations, although such a move would also affect Kaliningrad. The Baltic states have pursued alternate electricity supplies with the goals of diversifying options and reducing costs. For example, Lithuania connected to the Polish and Swedish electricity

---

[60] F. Stephen Larrabee, Stephanie Pezard, Andrew Radin, Nathan Chandler, Keith Crane, and Thomas S. Szayna, *Russia and the West After the Ukrainian Crisis: European Vulnerabilities to Russian Pressures*, Santa Monica, Calif.: RAND Corporation, RR-1305-A, 2017, p. 45.

[61] NATO Energy Security Centre of Excellence, homepage, undated.

[62] Kenneth Rapoz, "How Lithuania Is Kicking Russia to the Curb," *Forbes*, October 18, 2015.

[63] Latvijas Gāze, homepage, undated.

[64] Larrabee et al., 2017, pp. 30–33.

[65] Larrabee et al. 2017, p. 46.

markets in December 2015 and pursued but then abandoned plans with the other Baltic states for a joint nuclear power plant in January 2016.[66] Any development of an alternative grid for the Baltics would isolate Kaliningrad, which is already dependent on natural gas supplied through a pipeline in Lithuania, although this fact might not shift Russian decisionmaking.[67]

Economic Ties

Russia could also theoretically exploit its economic ties to the Baltic states. In 2013, one report noted, "As much as 19.8% of Lithuanian, 16.2% of Latvian and 11.4% of Estonian exports were directed to Russia in 2013."[68] The Baltic states share transportation infrastructure with Russia that dates from the Soviet period. For example, the Baltic countries have the same rail gauge as Russia, and a large majority of rail traffic through Estonia and Latvia are transshipments from Russia bound for Baltic ports.[69] Lithuanian analysts similarly highlight Lithuania's significant role in exporting goods from Europe to Russia—although Russia represents 20 percent of goods exported from Lithuania, only 4.8 percent of exports to Russia were from Lithuanian producers.[70] Transport companies in the Baltic states are disproportionately controlled by ethnic Russians. The Baltic countries also have been affected by Russian countersanctions of agricultural goods. The export of sprats, a sardine-like fish popular in both the Baltics and Russia, is

---

[66] Aivile Kropaite, "Baltic States Count Cost of Ending Soviet Electricity Link," *EU Observer*, December 15, 2015; discussions with Baltic official, Cambridge, United Kingdom, February 2016; "Baltic States Will Build New NPP In Lithuania," *The Baltic Review*, January 6, 2016.

[67] RAND MSW symposium, Cambridge, United Kingdom, February 3, 2016.

[68] Žygimantas Mauricas, "The Effect of Russian Economic Sanctions on Baltic States," Nordea Markets, undated.

[69] For example, in Estonia, an estimated 80 percent of rail traffic is transshipment. Olya Schaefer, "Estonian Transit Shows Growth but Fears Russia," Baltic Times, March 9, 2011; Paul Goble, "Moscow Launches Hybrid 'Rail War' Against Latvia," *Eurasia Daily Monitor*, Vol 12, No. 175, September 29, 2015b.

[70] Žygimantas Mauricas, "Lithuania: Economic Dependence of Russia," Nordea Markets, March 20, 2014.

commonly cited as the most visible victim of these sanctions.[71] The Baltics rank very low on Russia's list of trading partners—of the three countries, Latvia is the largest destination country, at 1.75 percent of exports, and Estonia is the largest exporter, at 0.35 percent of imports in 2016.[72]

Organized Crime and Criminal Networks

Following the end of the Soviet Union, organized criminal networks with ties to Russia and other former Soviet republics attempted to secure turf in the Baltic states. Tallinn "was ranked among the world's most violent capitals" in the early 1990s, but Estonia could quickly combat these groups and reduce their influence.[73] However, the Russian mafia had more success retaining connections and greater influence within Latvia, especially in the financial system. Partly because of its links with Russian organized crime, Latvia remained a major haven for money laundering.[74] Indeed, Mark Galeotti notes that although local gangs and law enforcement agencies in the Baltics were able to reduce the influence of Russian, Chechen, and Ukrainian gangs, Russian gangs "use the Baltic states for criminal services, ranging from money laundering through to gateways into Europe [sic]."[75] Other works have highlighted the connections between Russian organized crime and the intelligence services, implying that organized crime might be a significant opportunity for Russian influence in the Baltics.[76]

---

[71] Juris Kaža and Liis Kangsepp, "Baltic Countries Fear Impact of Russian Food Sanctions on Business," *Wall Street Journal*, August 7, 2014.

[72] Observatory of Economic Complexity, "Russia," webpage, undated.

[73] Jennifer Hanley-Giersch, "The Baltic States and the North Eastern European Criminal Hub," *ACAMS Today*, September–November 2009.

[74] Hanley-Giersch, 2009.

[75] Mark Galeotti, "Organized Crime in the Baltic States," *Baltic Review*, March 24, 2015.

[76] Brian Whitmore, "Organized Crime Is Now a Major Element of Russia Statecraft," *Business Insider*, October 27, 2015; Karen Dawisha, *Putin's Kleptocracy: Who Owns Russia?* New York: Simon and Schuster, 2014.

Cyber

The Baltic states, especially Estonia, are highly networked societies vulnerable to cyberattacks. Estonia has invested heavily in internet infrastructure for its government, not only for the sake of efficiency and cost, but also to maintain continuity of government in case the country is invaded. Estonia was the target of cyberattacks in 2007 around the Bronze Soldier crisis, as will be described in more detail in the next section. Although those attacks temporarily disabled some government and banking services, the overall impact on the country was limited—as one official put it, "disruptive but not destructive."[77] Partly as a result of the attack, Estonia invested significantly in cyberdefense, although Estonian officials remain concerned that foreign states will engage in cyberattacks in the future that will threaten Estonia's security.[78] Although Estonia has perhaps the greatest focus on the internet and cyber threat, malicious actors also pose significant threats to the other Baltic states' cyber infrastructures, a dynamic that led all three Baltic countries to sign a memorandum of understanding in November 2015 to promote cooperation in this area.[79] Estonia also hosts the NATO Cooperative Cyber Centre of Excellence.[80]

Means

Encouraging Ethnic Conflict

Through its compatriot policy, control over the media, use of information operations, and other government efforts, Russia can increase the tension in the Baltic states between the majority populations and the Russian-speaking or other minorities. As described in the previous discussion of Russia's strategy, Russia need not have a specific end state

[77] Phone call with Estonian official, November 2015.

[78] See KAPO, Annual Review, 2014, p. 18.

[79] "Memorandum of Understanding (MoU) Between Ministry of Economic Affairs and Communications of the Republic of Estonia and Ministry of Defence of the Republic of Latvia and Ministry of National Defence of the Republic of Lithuania," November 4, 2015.

[80] NATO Cooperative Cyber Defence Centre of Excellence, homepage, undated.

in mind to encourage ethnic conflict. Such conflict in the Baltics, for example, could undermine those countries' efforts to build sustainable, multiethnic democracies, thereby challenging the Western goal of greater Euro-Atlantic integration. It could also reduce NATO's support for the Baltics, making them more vulnerable to further subversion. Ethnic conflict might also facilitate an increase in political power for individuals over whom Russian could exert greater influence. Finally, ethnic conflict could escalate to violence or riots, the way it did in Estonia in 2007, creating a situation that Russia could exploit to its advantage. Based on Russia's extensive ties with the Russian speakers in Estonia and Latvia, these are the main—and, perhaps, most likely—minorities that Russia would support. But Moscow could also support other ethnic minorities, such as the Poles in Lithuania.

Analysts highlight early Russian intentions to encourage ethnic conflict in the Baltics. In 1992, Sergei Karaganov, a Russian analyst, wrote an article for a magazine published by the Russian Ministry of Foreign Affairs in which he advocated keeping the Russian-speaking population in the Baltics while acting to prevent their integration. This policy of maintaining the Russian speakers as a separate group and a tool of influence came to be referred to as the *Karaganov doctrine*.[81] Following the 2004 Orange Revolution in Ukraine, Russian officials began to interpret their policies in the near abroad as a failure. They developed a new policy and committed additional resources for the near abroad that became codified in Russia's compatriot policy.[82]

The *compatriot policy* is executed by several governmental and semigovernmental organizations that are supported by other Russian government organizations. The Russian Ministry of Foreign Affairs contains a Department for Cooperation with Compatriots, which works closely with *Rossotrudnichestvo*, an agency that is subordinated to the ministry.[83] *Rossotrudnichestvo* works with and provides money to several NGOs, including, most prominently, the *Russkiy Mir* foundation. These orga-

---

[81] Winnerstig, 2014, pp. 113–114.

[82] Laruelle, 2015, pp. 9–10.

[83] Federal Agency for the Commonwealth of Independent States, Compatriots Living Abroad, and International Humanitarian Cooperation, undated.

nizations then provide money, advice, and support to local pro-Russian organizations in the Baltics.[84] Indeed, Estonian and Latvian organizations emphasize Russia's funding and coordination of these groups' activities to demonstrate Russian influence.[85] In Latvia, for example, a 2012 report uncovered that more than 20 organizations had received money from the *Russkiy Mir* foundation, contrary to Latvian law requiring disclosure of such funding.[86] Think tank reports have also alleged Russian intelligence work in coordination with some of these organizations.[87]

Russia's compatriot policy is likely tied to the development of pro-Russian organizations operating within the Baltics as already discussed, including those associated with *Mir Bez Natsizma*.[88] Estonian sources claim that there were close consultations between the Russian

---

[84] *Russkiy Mir*, homepage, undated. The Estonian KAPO noted "a new government-financed fund, Istoria Otechestva (History of the Fatherland), was published. This fund will be given the task of introducing the history of Russia at home and abroad, and to support history education programmes. The media has opined that the new fund's activity will be similar to that of another Russian government fund, Russkii Mir, which supports the segregation of Russian expatriates." KAPO, 2014, p. 11. See also Marcel van Herpen, *Putin's Propaganda Machine: Soft Power and Russian Foreign Policy*, Lanham, Md.: Rowman and Littlefield, 2016, p. 36.

[85] For example, Pelnens (2009) notes about Latvia,

> Russian interests are also looked after by the Moscow House in Riga, founded and run by the government of Moscow and envisaged "for humanitarian and business" partnerships with Russian compatriots residing abroad. Russian officials habitually see organizations of Latvia's Russian speakers as their natural partner for disseminating information, organizing seminars or conferences, recruiting participants for mass rallies and pickets, and collecting signatures for petitions to international institutions and EU governments claiming discrimination against national minorities. (pp. 156–157)

[86] Other accounts suggest the existence of more than 100 Russian-funded organizations. Paul Goble, "Moscow Using Russian Organizations to Destabilize Latvia, Riga Officials Say," *The Interpreter*, March 10, 2015.

[87] In Estonia, for example, Pelnens (2009) noted, "Vladimir Pozdorovkin, who worked at the S.R.V.'s Political Intelligence Central Administrative Board, participated as a patron in a compatriots conference held in June of 2007, introducing to its participants the Russkiy Mir (Russian World) compatriots program" (p. 70).

[88] See KAPO, 2014, p. 6; Latvia Security Police, "Annual Report," 2013, p. 12.

embassy and the Night Watch, a pro-Russian Estonian organization that played a significant role in the 2007 protests.[89]

Indeed, the details of the 2007 Bronze Soldier crisis offers a means to understand how Russia has encouraged mobilization of the Russian-speaking minority and how it might do so again.[90] In its campaign of the 2007 election, the Estonian Reform Party, led by Andrus Ansip, "used the monument's removal as a means of mobilizing support for itself" among ethnic Estonians.[91] The party was elected and initiated a process of moving the Bronze Soldier statue, which was a memorial to Soviet troops who died fighting in World War II. However, Russian speakers tended to see the decision to move the statue as a symbol of Estonia's rejection of Soviet history, with echoes of collaboration with the Nazi

---

[89] For example, the International Center for Defense and Security notes,

> There are grounds to suspect that the Embassy of the Russian Federation in Tallinn, the capital of Estonia, has been directly instructing local extremists and organizers of unrest. According to the Estonian Security Police, during the weeks leading up to the disturbances that took place in Estonia, Senior Counselor of the Embassy of the Russian Federation Sergei Overtshenko met repeatedly at the Tallinn Botanical Gardens with Dmitri Linter, who is the leader of "The Night Patrol"—the grouping that is suspected of having organized the rioting.

International Center for Defense and Security, "Russia's Involvement in the Tallinn Disturbances," May 11, 2007. See also Kadri Kukk, "Brief History of 'Night Watch' in Estonia," Café Babel, May 7, 2007. Nevertheless, an Estonian court acquitted the Night Watch of charges of riot instigation. "Estonian Court Acquits Defendants in Bronze Soldier Protests," Sputnik International, May 1, 2009.

[90] Drawn from unpublished research by Stephanie Pezard, Katya Migacheva, and Brenna Allen, as well as Andrew Radin and Katya Migacheva. We thank these authors for sharing this material.

[91] Kaiser quotes Ansip as explaining:

> I see the solution to this problem in the relocation of the monument to the cemetery. . . . It has become all the more clear that the monument cannot remain in its old place. The question rose: whose word has authority in Estonia? The word coming from the Kremlin or the word from Old Town? We cannot say to our people, that Estonia is after all only a union republic, and our word in this country is not worth a 'brass farthing.'

Robert Kaiser, "Reassembling the Event: Estonia's 'Bronze Night,'" *Environment and Planning D: Society and Space*, Vol. 30, 2012, pp. 1051–1052.

regime.[92] Soon after the monument was fenced off on April 26, 2007, a crowd estimated at approximately 1,000 and mainly composed of ethnic Russians formed to protest the movement of the statue. That evening, the police attempted to disperse the crowd, resulting in violence, looting, and one death.[93]

Estonian sources highlight the Russian role in the events. Merle Maigre, a former adviser to the Estonian president, refers to the "Bronze Night" as "a conflict of a hybrid nature," including "riots in Tallinn, a siege of the Estonian Embassy in Moscow by pro-Kremlin *Nashi* youth organization demonstrators, strong economic measures imposed by Russia against Estonia, waves of cyberattacks against the Estonian government and banking systems, and a fiery official Russian response."[94] Similarly, a well-respected Estonian think tank, the International Center for Defense and Security, published an article claiming that the Russian embassy directly participated in the organization of the protests and highlighting Russian support for the protests after the fact, including through sanctions, cyberattacks, and encouragement of the Russian *Nashi* youth group to organize protests around the Estonian embassy in Moscow.[95]

The events of April 2007 are indicative that Russia's role was encouraging pro-Russian sentiment that made the crisis possible and supporting the protests once they began, not fully orchestrating the development of the crisis. Prior to the protest, Russia invested resources in encouraging the adoption of its own narrative of World War II and likely provided financial support to some pro-Russian groups, including the Night Watch, that played a role in organizing the protests. The speed with which the protests emerged, however, is inconsistent

---

[92] For example, Russian media coverage highlighted an effort by Estonian veterans' groups to erect a statue commemorating the Estonian SS legion. Karsten Brüggemann and Andres Kasekamp, "The Politics of History and the 'War of Monuments' in Estonia," *Nationalities Papers*, Vol. 36, No. 3, 2008, p. 448.

[93] Brüggemann and Kasekamp, 2008, p. 436.

[94] Merle Maigre, "Nothing New in Hybrid Warfare," German Marshall Fund of the United States, policy brief, February 12, 2015, p. 4.

[95] International Center for Defense and Security, 2007.

with prior planning from Moscow. The presence of cyberattacks that were apparently perpetrated and organized using the Russian language is also taken as a sign of Russian involvement, but the timing and sophistication of these attacks is also inconsistent with prior support from Moscow. A relatively unsophisticated first wave of attacks occurred from April 27 to April 29, including use of simple scripts to ping Estonian servers. One report described these attacks as "emotionally motivated, as the attacks were relatively simple and any coordination mainly occurred on an ad hoc basis."[96] A second wave of attacks that was far more sophisticated occurred from April 30 to May 18, including the use of botnets to conduct distributed denial-of-service attacks on Estonian banks and official websites. Still, according to a U.S. analyst at the time, the attacks "from a technical standpoint [are] not something we would consider significant in scale."[97] These events are more consistent with post hoc Russian support, an interpretation agreed with by U.S. and Baltic analysts.[98] The events of 2007 imply that although Russia might encourage ethnic tensions that lead to riots or protests, it is less likely to intentionally organize pro-Russian protests from scratch.

Russia could also encourage ethnic conflict in Lithuania, although in the absence of large numbers of Russian speakers, the environment is likely less permissive. One report by Lithuanian authors quotes Dugin's encouragement to exacerbate ethnic tensions in Lithuania: "Ethnic tensions between Lithuanians and Poles are an especially valuable asset and should be used, or, whenever possible, these tensions

[96] The initial attack apparently began when instructions for executing ping commands were posted on various Russian-language internet sites. One paper explains, "As a generalisation, though, the initial attacks on April 27 and 28 were simple, ineptly coordinated and easily mitigated." Eneken Tikk, Kadri Kaska, and Liis Vihul, *International Cyber Incidents: Legal Considerations*, Talinn, Estonia: Cooperative Cyber Defence Centre of Excellence, 2010, p. 18.

[97] Quoted in Roland Heickerö, *Emerging Cyber Threats and Russian Views on Information Warfare and Information Operations*, Stockholm, Sweden: FOI Swedish Defense Research Agency, 2010, p. 42.

[98] Discussions with U.S. and Baltic analysts and officials, by phone and in Washington D.C., and Cambridge, United Kingdom, November 2015–April 2016.

should be deepened."[99] Although Russia does appear to have invested in the development of pro-Russian organizations, the *Russkiy Mir* and other organizations do not appear to have much of a presence in Lithuania.[100] Although there might be potential for Russia encouraging Polish- and Russian-speaking parties in the future, possibly in combination with other tools of influence, Lithuanian officials downplayed concerns about direct Russian influence through ethnic minorities.[101]

Nonethnic Political Manipulation

Russia has other means to influence the politics of the Baltic states besides creating ethnic tensions. One such tool is to support policies or legislation that would divide the Baltic states from the EU and/or lead to greater political or ideological alignment with Moscow. For example, one Latvian analyst noted that Russia has supported conservative legislation on such issues as same-sex marriage, hoping to build a coalition with conservative parties in Latvia.[102] In doing so, Russia hopes to identify and make more salient the issues that would align the Baltic countries with itself rather than with the EU and the United States. Still, memory of occupation by the Soviet Union is probably more important in the Baltic public discourse than conservative values shared with Russia—especially because, as Latvian analysts emphasize, Russia's adherence to conservative values is inconsistent.[103]

Furthermore, Russia might co-opt officials by promoting personal corruption. In one discussion, it was suggested that Russia might buy off a member of the ruling coalition in Estonia or Latvia to pave the way for a Russian majority party to enter office.[104] The presence of Russian organized crime also might facilitate personal corruption.

---

[99] Winnerstig, 2014, p. 118

[100] Winnerstig, 2014, pp. 126–127.

[101] RAND MSW symposium, Cambridge, United Kingdom, February 2, 2016

[102] Phone call with Latvian analyst, December 2015; Michael Birnbaum, "Gay Rights in Eastern Europe: A New Battleground for Russia and the West," *Washington Post*, July 25, 2015.

[103] Discussions with Latvian analysts, by phone and in Riga, July and December 2015.

[104] Discussion with Latvian analyst, Riga, July 2015.

However, the well-developed security forces and legal systems of the Baltic states likely pose a strong deterrent to corruption compared with other former Soviet republics.[105]

Another angle to bear in mind is that Russian propaganda and information operations might have a more pervasive influence that extends beyond merely Russian speakers in the Baltics. Lithuanian analysts highlight the popularity within Lithuania of Russian media—especially television because of higher-quality programming. A central concern is that Russia will be able to spread its narrative about World War II that depicts the Baltic governments as the successors of fascism, thus undermining the independence of the Baltic states and harming their integration into Western institutions.[106]

"Little Green Men"—Covert or Deniable Action

Following events in Crimea and eastern Ukraine, analysts have expressed growing concern that Russia would replicate similar strategies of using special forces, intelligence, or other covert or deniable means in the Baltic states. Baltic officials and NATO analysts emphasize that it is unlikely that Russia's actions in the Baltic states would precisely replicate the actions it took in Ukraine both because the environment in the Baltics is different and because there are significant questions regarding the extent to which operations in Ukraine succeeded.[107] Furthermore, there are questions about what might motivate

---

[105] For example, one report notes Latvia's improving rule-of-law institutions despite the extensive economic ties between Russia and Latvia likely make Latvia less susceptible to Russian organized crime or corruption activities. Heather Conley, James Mina, Ruslan Stefanov, and Martin Vladmirov, *The Kremlin Playbook: Understanding Russian Influence in Central and Eastern Europe*, Lanham, Md.: CSIS Europe Program and CSD Economics Program, Rowman and Littlefield, October 2016, pp. 49–50.

[106] One report notes that there are "Russian television channels on Lithuanian cable networks and Lithuanian television channels that are overflowing with Russian productions" (Winnerstig, 2014, p. 129). Furthermore, "a large portion of the population receives not just entertainment, but also news about the world and the post-Soviet region through the Russian media" (Winnerstig, 2014, pp. 131–132). See also Pelnens, 2009, pp. 197–200.

[107] Samuel Charap, "The Ghost of Hybrid War," *Survival*, Vol. 57, No. 6, December 2015, pp. 53–55; Ruslan Pukhov, "Nothing 'Hybrid' About Russia's War in Ukraine," *Moscow Times*, May 27, 2015; discussions with Baltic defense officials, Tallinn and Riga, July 2015.

Russia to take such an aggressive action. The posited reason for a Russian incursion in the Baltics would be to roll back NATO to a region of lesser interests rather than to stop NATO enlargement in a region of higher interests.[108] Nevertheless, the potential for Russian "little green men" to undermine NATO in the Baltics is clearly on the minds of many NATO and U.S. policymakers.

Russia's operations in Ukraine are the most recent demonstration of Russia's tactics, techniques, and procedures for using covert military force. These operations have been examined in detail elsewhere, but it is worth identifying two different strategies that shed light on how Russia might approach conflict in the Baltics. First, Russia used snap exercises to launch an operation by Spetsnaz and other ground forces to rapidly seize Crimea, supported by information operations, intelligence, and other techniques. Russia then annexed and consolidated control over the territory. Second, Russia's operations in eastern Ukraine had a quite distinctive character and evolved over the course of 2014. It appears that Russia initially used proxies and intelligence assets to shift the trajectory of protesters supporting Ukraine President Viktor Yanukovitch toward separatism. In the spring of 2014, the separatists could seize territory and establish themselves because of the weakness of the Ukrainian military. In July of the same year, Ukraine launched a more substantial attack on the separatists, which forced Russia to respond with greater support for the separatists, as demonstrated by use of a Russian-produced anti-aircraft weapon to shoot down Malaysian Airlines flight MH-17. With the separatists on the verge of defeat, Russia engaged with conventional forces, including firing artillery from over the border and deploying Russian ground forces into Ukraine.[109]

---

[108] Radin, 2017b.

[109] For detailed analysis, see Andrew Radin, *Hybrid Warfare in the Baltics: Threats and Potential Responses*, Santa Monica, Calif.: RAND Corporation, RR-1577-AF, 2017a; Charap, 2015; Michael Kofman, "Russian Hybrid Warfare and Other Dark Arts," *War on the Rocks*, March 11, 2016; Charles K. Bartles and Roger N. McDermott, "Russia's Military Operation in Crimea," *Problems of Post-Communism*, Vol. 61, No. 6, 2015; Michael Kofman, Katya Migacheva, Brian Nichiporuk, Andrew Radin, Olesya Tkacheva, and Jenny Oberholtzer, *Lessons from Russia' Operations in Crimea and Eastern Ukraine*, Santa Monica, Calif.: RAND Corporation, RR-1498-A, 2017.

Russia's actions in Ukraine, and discussions with officials and analysts, point to three potential means by which Russia could use covert military action in the Baltics. First, following the model of Crimea, unmarked Russian forces could enter Russian-dominated areas, such as Narva in Estonia or Latgale in Latvia, in a large planned operation. This operation could happen after efforts at political mobilization by Russia to encourage a separatist movement or other opposition by local Russians. After Russian forces helped seize the area, separatist leaders might declare their intent to secede and be annexed into Russia, which would make Russia's intention to defend the area with nuclear weapons more credible. Russian actions might be covert—or, at the very least, denied—in an effort to limit the potential for a strong NATO response before the territory could be annexed. In a second scenario, following the model of eastern Ukraine in the summer of 2014, Russia could offer covert support for a separatist movement in Russian-dominated areas without deploying significant forces. Under this scenario, the nascent separatist movement might not be able to control territory in the Baltics, but it would be easier for Russia to deny its involvement. Third, Russia could attempt to encourage a terrorist campaign or other violent insurgency throughout the territory of the Baltic states.[110]

There are several challenges to such operations. First, Russia would rely on significant support from local Russian speakers in the Baltics. As this discussion emphasizes, Russia would have difficulty recruiting large numbers of supporters or sustaining support for a separatist movement that would justify and legitimate Russian involvement. Second, the Baltic countries have plans to rapidly defeat pro-Russian forces and force Russia to escalate the conflict or accept the defeat of its proxies. The Estonian Chief of Defense, General Riho Terras, noted, "If Russian agents or special forces enter Estonian territory, you should shoot the first one to appear. . . . If somebody without any military insignia commits terrorist attacks in your country

---

[110] Radin, 2017a, p. 24.

you should shoot him. . . . [Y]ou should not allow them to enter."[111] Latvian officials have apparently made similar comments.[112] Russia's response to the defeat of its proxies in Ukraine was to escalate to conventional warfare.[113] However, NATO's Article 5 guarantee commits NATO member states to treat an attack on the Baltic states as an attack on themselves, which could deter Russia from escalating to the overt use of military power. The Baltic countries' overall strategy for deterring the "little green men" is to engage in a significant campaign to defeat irregular Russian forces early in the conflict.

This strategy leads to the question of the extent to which the Baltic countries can defeat irregular, Russian-backed forces.[114] The Baltic countries, although small, have modern capabilities and substantial reserves. They exercise greater control over their own territory than Ukraine, for example, and academic research on insurgency and rebellion emphasizes the difficulty of sustaining a rebellion where the opposing government exercises control. With territorial control, it becomes easier for the state to gather information about the rebel group, target opponents, and encourage defection.[115] This appears to be especially true when governments are fighting a foreign-supported rebellion; for-

---

[111] Sam Jones, "Estonia Ready to Deal with Little Green Men," *Financial Times*, May 13, 2015. Terrass also said that

> "Hybrid warfare is nothing new. You can deal with it only with the cohesion of the nation, with integrity, with all society working together. . . . [Estonia] is a functioning society," he stressed. "We are not like Ukraine. . . . But we need to be very well aware of what is happening in Russia and be ready." Most importantly, Gen Terras said, [NATO] needed to be prepared to stand behind his country and go to war in the event of his forces having to forcibly confront any Russian interference in a way that Kiev was initially unable to do.

[112] Interviews with Latvian defense officials, Riga, July 2015.

[113] Charap, 2015.

[114] RAND research has demonstrated that NATO would be comprehensively outmatched in a short-notice conventional conflict in the Baltics given current posture, although NATO possesses superior conventional forces to Russia overall. David A. Shlapak and Michael W. Johnson, *Reinforcing Deterrence on NATO's Eastern Flank: Wargaming the Defense of the Baltics*, Santa Monica, Calif.: RAND Corporation, RR-1253-A, June 2015.

[115] For example, see Stathis Kalyvas, *Logic of Violence in Civil Wars*, New York: Cambridge University Press, 2006.

eigners will have more difficulty dissuading locals from informing on them and finding locations to hide among the local population.

Further research needs to be done to understand how the Baltic countries would fare against well-trained Russian special forces or local militias backed by Russian forces and to understand how the Baltic forces might adopt asymmetric tactics to fight a conventional Russian force.[116] Russia also might be able to leverage other hostile measures to support military action in the Baltics. For example, it may use cyber warfare to disrupt Baltic countries' information systems and undermine their responses; or it might use information operations, proxies, or other means of influence to dissuade a NATO response. Although the deployment of U.S. or NATO troops might impose difficulties on the operations of Russian irregulars or Russian proxies, defense against cyber or other hostile measures might require other forms of NATO assistance. Nevertheless, the ability of even the unprepared and depleted Ukrainian military to fight the Russian-backed separatists in July and early August of 2014 indicates that the Baltic countries would have a good chance of pushing back and defeating Russian-backed irregular forces.

Economic Leverage

Another possible Russian means of influence is the threat of cutting off trade or imposing other types of sanctions to harm the Baltic states or compel them to adopt a particular policy.

Russia has imposed economic sanctions on the Baltic states in the past. For example, following the Bronze Soldier incident, the Russian Rail Company ceased oil deliveries to Estonia, citing the need for repairs,[117] and some Russian companies refused to buy Estonian products. One newspaper in Russia identified a 3-percent loss in GDP resulting from

---

[116] Jan Osburg, *Unconventional Options for the Defense of the Baltic States: The Swiss Approach*, Santa Monica, Calif.: RAND Corporation, PE-179-RC, 2016.

[117] "Ссора с россией обернулась для эстонии огромными потерями [Fight with Russia Brought Estonia Great Losses]," zagolovki.ru, November 17, 2007; Nikolai Starikov, "Почему эстонцы ведут себя так нагло, а Россия так сдержанно? [Why Are Estonians Behaving So Boldly and Russia So Timidly?]," *Internet vs. TV Screen*, undated.

Russian economic sanctions.[118] The Baltics have also been affected since 2014 by the sanctions and countersanctions associated with Russia's actions in Ukraine. One article estimates the value of sanctioned goods at "2.6% of GDP in Lithuania, 0.4% of GDP in Estonia, and 0.3% of GDP in Latvia."[119] However, these numbers include re-export of goods through the Baltic states. When this is considered, the impact is estimated to be significantly lower: 0.4–0.5 percent of GDP in Lithuania, 0.2–0.3 percent in Estonia, and 0.1–0.2 percent in Latvia.[120]

In general, the Baltic countries might not be especially vulnerable to Russian economic coercion. When asked whether the Baltic countries would be affected by Russia's decision to cut off imports, Baltic officials observed that the companies involved understood the risks and downplayed the potential political impact of such a decision.[121] One telling example is the Baltic countries' effort to shift away from the Russian rail gauge to the European standard. This decision could come with a significant economic cost from the loss of Russian traffic, although Russia is developing its own ports, and traffic to Estonia might decrease in any case.[122] Nevertheless, there appears to be significant political support within the Baltics for efforts to decrease economic dependence on Russia, risking short-term economic cost in favor of greater long-term trade with Europe.

Russian sanctions, or the threat of them, have not always been associated with a political goal. In the case of Russia's threat of cutting off trade with Ukraine, there was a specific goal of persuading Yanukovitch not to sign the Deep and Comprehensive Free Trade Agreement with the EU.[123] In other circumstances, Russia might not make specific

---

[118] Maxim Lensky and Nikolai Flichenko, "Tallinn: A Year Without the Bronze Soldier," *Kommersant*, April 25, 2008.

[119] Kaspar Oja, "No Milk for the Bear: The Impact on the Baltic States of Russia's Counter-Sanctions," *Baltic Journal of Economics*, Vol. 15, No. 1, 2015, p. 38.

[120] Oja, 2015, p. 46.

[121] Discussions with Estonian, Latvian, and Western officials, Tallinn and Riga, July 2015.

[122] Schaefer, 2011.

[123] David Herszenhorn, "Facing Russian Threat, Ukraine Halts Plans for Deals with E.U.," *New York Times*, November 21, 2013; "How the EU Lost Russia over Ukraine," Spiegel

or realistic demands. In the case of the Bronze Soldier crisis, for example, Russian sanctions could have been intended more for a domestic audience to demonstrate the government's frustration with the treatment of Russian speakers in Estonia. Alternatively, Russia might have sought to undermine Estonia's resistance or opposition without a specific policy goal in mind. Sanctions might also have domestic economic benefits: In the case of countersanctions related to the Ukraine conflict, Russia likely sought the end of Western sanctions and used the countersanctions to strengthen Russia's agricultural sector.[124]

Russian economic influence is largely exercised through Russian companies. Although Russian companies might in part be a means of political influence, their activities might be justified and based on economic motives. For example, energy company Gazprom's policies in shutting off gas supplies to Ukraine in 2007 and 2009 were influenced both by Russia's policy that Ukraine would no longer receive significant discounts on natural gas following the Orange Revolution and by the goal of ensuring that Naftogaz Ukrainy paid for Ukraine's gas consumption.[125] Although Russia had a number of political grievances against Ukraine in 2014 and apparently did raise the threat of increasing gas prices to pressure Yanukovitch not to sign the Deep and Comprehensive Free Trade Agreement, the failure of Ukraine to pay the agreed-upon price likely also contributed (again) to the decision to cut off gas supplies.[126] Although the activities of Russian companies motivated by profit might not initially appear to fall under the rubric of hostile measures (because such activities are not intended for political influence), Russian policymakers might see no such distinction. For them, the economic fortunes of Russian companies are closely connected to the success of the

---

Online International, November 24, 2014.

[124] "Agriculture Minister: Russian Food Will Squeeze Out Imports in 10 Years," *Moscow Times*, July 7, 2015.

[125] Abdelal, 2013.

[126] Discussions with Russian and U.S. analysts, Washington D.C., April 2016; Paul Kirby, "Russia's Gas Fight with Ukraine," BBC News, October 31, 2014.

Russian state and vice versa.[127] To understand, combat, and respond to Russian economic coercion, Western policymakers should recognize the interlinked political and economic motives and activities of the Russian state and Russian companies. Russian economic coercion in the Baltics or elsewhere, for example, might be limited by concerns about damaging the economic prospects of Russian companies elsewhere in Europe.

Bullying with Military and/or Intelligence Forces

A final means of Russian influence is simple bullying—intimidating or threatening the Baltic states with military or intelligence forces. Bullying need not have a specific purpose in mind; indeed, it may simply be intended to remind the Baltic states of Russia's superior military capabilities and ability to cause harm if it decides to do so.

There are several examples of Russian bullying.[128] First, Russia has intensified the number of overflights of Baltic territories. A report in

---

[127] For example, Fiona Hill and Clifford Gady write,

> In sum, the importance of oil and gas as the main sources of value for Russia, and the significance of transportation infrastructure as a means of ensuring control over the physical flows of oil and gas, have helped Putin to define which companies needed to be in the core of his Russia, Inc. In this context, the juridical ownership of the core Russian companies has proven almost irrelevant. . . . In all cases, Mr. Putin. . . has seen to it that these companies are inside his system and have been subject to his oversight in both formal and informal ways.

Fiona Hill and Clifford Gaddy, *Mr. Putin: Operative in the Kremlin*, Washington, D.C.: Brookings Institution, 2015, p. 226.

[128] See also Elisabeth Braw, "Bully in the Baltics: The Kremlin's Provocations," *World Affairs Journal*, March/April 2015c. Stephen Covington describes Russian intimidation as part of an effort to break out of their perceived "strategic encirclement" within the current European security system. He notes that

> Russia's breakout actions include the use of force in Crimea, withdrawal from the CFE treaty, military, financial, and political support to separatists in eastern Ukraine, direct financial, political, and military actions to destabilize Ukraine on a broader scale, a military rearmament program, the buildup of military capabilities in the Arctic, Black Sea, and Baltic Sea, sudden large-scale military exercises that shift forces to higher combat readiness involving long-range deployments, nuclear force exercises designed to posture and intimidate, and energy, financial, and informational pressure on European countries. All of these political and military actions break with the norms, rules, and practices of the post-Cold War period and destabilize the current security system.

November 2014 highlighted a significant increase in Russian aircraft approaching the territory of the Baltic countries and hypothesized that these activities go beyond training purposes to "serve as a demonstration of Russia's capability to effectively use force for intimidation and coercion, particularly against its immediate neighbors."[129]

Snap exercises are a second means of intimidation. Russia's invasion of Crimea was launched in part from a snap exercise, so a precedent exists for Russia to use an exercise as a jumping-off point for a major military action. Ieva Berzina, a Latvian analyst, describes a major Russian exercise, Zapad 2013, as a "form of strategic communication" and notes that, from the perspective of the Baltic officials, the exercise "destabilizes the security environment of the Baltic states." [130] Berzina claims that "from the perspective of Russian strategy, the alarming reactions of the Baltic states and Poland was the necessary effect, because the intimidation of an adversary is a very important element of the strategy of deterrence."[131] Whatever the connection between deterrence and the Russian exercises, it is likely that Russia sought to produce feelings of insecurity within the Baltic states.

Third, Russian intelligence has engaged in several aggressive activities against the Baltic states and could continue to do so. Prominent examples to date include Vladimir Veitman, a KAPO officer who was

---

Stephen Covington, *Putin's Choice for Russia*, Cambridge, Mass.: Belfer Center for Science and International Affairs, August 2015, p. 10.

[129] Thomas Frear, Łukasz Kulesa, and Ian Kearns, *Dangerous Brinkmanship: Close Military Encounters Between Russia and the West in 2014*, London: European Leadership Network, November 2014, p. 2. The report also noted,

> NATO officials indicated in late October 2014 that this year NATO states have already conducted over 100 intercepts of Russian aircraft, three times more than in 2013. (1) Between January and September, the NATO Air Policing Mission conducted 68 "hot" identification and interdiction missions along the Lithuanian border alone, and Latvia recorded more than 150 incidents of Russian planes approaching its airspace. (2) Estonia recorded 6 violations of its airspace in 2014, as compared to 7 violations overall for the entire period between 2006 and 2013. (p.10)

[130] Ieva Berzina, "Zapad 2013 as a Form of Strategic Communication," in Liudas Zdavavičius and Matthew Czekaj, eds., *Russia's Zapad 2013 Military Exercise: Lessons for Baltic Regional Security*, Washington, D.C.: Jamestown Foundation, 2015, p. 70.

[131] Berzina, 2015, p. 70.

revealed in 2013 to have been recruited by the foreign intelligence service; Herman Simm, who was caught in 2009 for sharing NATO classified information with Russia; and Eston Kohver, the Estonian internal security officer who was captured by the FSB on the Russian border in September 2014.[132] Some analysts question whether the capture of Kohver was approved by Russian central leadership, given that he appeared to be of little strategic value and that Moscow eventually exchanged him for a detained Russian agent.[133] Regardless, Russian intelligence agencies maintain a significant presence within the Baltic countries.[134]

These activities might seem unlikely to have the effect of persuading the Baltic states to follow Russia's leadership more closely. In many ways, such activities are more likely to convince the Baltic states and NATO that a greater Western military and intelligence presence is required to counter or deter Russian actions. Still, such actions are in line with a description of Putin, and of Russia's leaders in general, as having the mentality of KGB agents or criminal gangs acting as bullies, hoping to use their greater leverage and military capability to gain strategic advantage.[135]

---

[132] See Fidelius Schmid and Andreas Ulrich, "Betrayer and Betrayed: New Documents Reveal Truth on NATO's 'Most Damaging' Spy," Spiegel Online International, May 6, 2010; "Veitman Sentenced to 15 Years for Spying for Russia," Estonian Public Broadcasting, October 30, 2013; KAPO, 2014, pp. 2–3.

[133] Discussions with Estonian officials and United Kingdom analysts, Tallinn and London, July 2015, February 2016; Argo Ideon, "Expert: FSB Likely Fearing Capture of Vital Agent in Estonia," *Postimees*, December 30, 2014.

[134] Weiss, 2014. For example, Kohver was exchanged for Aleksei Dressen, a KAPO officer convicted of working for the FSB, and in 2014, Russia also publicized statements from Uno Puusepp, another former KAPO officer who claimed to be working for Russia. Ideon, 2016; "FSB Reveals It Had Agent in Estonia Intel for 20 Years," *Rossiyskaya Gazeta* via Interfax, December 14, 2014.

[135] Hill, 2016; RAND MSW symposium, Cambridge, United Kingdom, February 2, 2016; discussions with analysts, Washington D.C., April 2016.

The Future of Russian Hostile Measures in the Baltics

The Baltics might seem at first glance like the greatest point of vulnerability in NATO and the EU to Russian hostile measures because of their proximity to Russia, the Russian-speaking minorities in Estonia and Latvia, and longstanding economic ties. However, Russia's expressed priorities and our assessment in Chapter One offer good reasons to think that the region will not be a priority for Russian foreign policy. Compared with Ukraine or other former Soviet republics, Russia's hostile measures do not appear to be especially intensive.[136] Nevertheless, Russia could increase its efforts to undermine the stability of the Baltic states through the measures we have discussed, including information operations, engagement with the Russian-speaking populations in Estonia and Latvia, covert action, or cyberattacks.

It is likely that the most effective means of Russian subversion is working through Russian-speaking communities. Russia has invested some resources in maintaining influence in the Baltics through its compatriot policy, and it has had some success in helping pro-Russian organizations there. It will likely continue this low-cost policy in the future, hoping to retain some degree of influence and to leave open the possibility of provoking crisis, especially on ethnic issues, that could lead to future opportunities. However, Russian-speaking populations will not be easy to manipulate or to use to undermine the countries' stability because of the integration and loyalty of many Russian speakers to their home countries.

Other hostile measures appear unlikely or ineffective. Although the Baltics do have substantial trade with Russia and some degree of energy independence, this dependence is in decline and there is strong political will not to accede to Russian demands. Covert military action remains a concern, although the Baltic countries, bolstered by a conventional NATO force, can likely counter this threat short of a major Russian military operation. Russia will likely continue to use mili-

---

[136] See, for example, Linda Robinson, Todd C. Helmus, Raphael S. Cohen, Alireza Nader, Andrew Radin, Madeline Magnuson, Katya Migacheva, *Modern Political Warfare: Current Practices and Possible Responses*, Santa Monica, Calif.: RAND Corporation, RR-1772-A, 2018; Leonhard and Phillips, undated.

tary means to bully the Baltics, but it remains perhaps more likely that these actions will reinforce the region's adversarial policy toward Russia rather than weakening it. Although corruption or other types of nonethnic political manipulation could have a dramatic effect, the Baltic states' recent political stability and commitment to the rule of law suggest that Russian-sponsored corruption would likely have little systemic effect on the countries' stability.

Manipulation of Ethnic Conflict

To address the biggest vulnerability of Russian subversion through the Russian-speaking community, one option is to encourage the Baltic countries to revisit the status and citizenship of Russian speakers and reduce their grievances. These issues are extremely politically sensitive in the Baltic countries—one analyst explained his belief that if Russian were accepted as an official language in Latvia, Latvian would disappear from use within two generations.[137] Since 1991, Estonia and Latvia have seen the current policies toward Russian speakers as essential for developing their countries as independent states linked to the titular national group, and they will likely retain this belief in the indefinite future. Because these countries are full members of the EU and NATO, there are few obvious levers to persuade them to reconsider these beliefs.

A second approach would be to focus on Russian-language strategic communication. The dominance of Russian media in the Russian speakers' information space is troubling. Although there have been efforts to develop alternative Russian-language media in Estonia, efforts in Latvia have been less successful. The U.S. military should not necessarily be in the lead on strategic communications in the region, but it might have unique capabilities and a role to play in developing and deploying appropriate communications with Russian speakers.[138]

---

[137] Interview with Latvian analyst, Riga, July 2015.

[138] See also Helmus et al., 2018.

U.S. and NATO Military Engagement

Since 2016, efforts to strengthen the Baltic states against possible Russian aggression have intensified, including through the provision of forward military presence to deter a possible Russian attack. Among other deployments, NATO has deployed four enhanced forward presence battalions to the Baltic states and Poland, and the United States has provided a rotational armored brigade combat team in Poland.[139] A wide range of U.S. bilateral engagement and assistance has also been provided to the Baltic states, including through the European Deterrence Imitative (previously known as the European Reassurance Initiative). Examples include training for joint terminal attack control and forward air controller interoperability; border security; intelligence, surveillance, and reconnaissance (ISR) capabilities; and improving local infrastructure.[140]

In assessing their priorities for U.S. military assistance, Baltic state officials tend to rank the threat of Russian hostile measures below the potential conventional military threat posed by Russia. Estonian officials, for example, do not reject possible U.S. assistance on hostile measures but emphasize their capabilities on cyber warfare and the domestic responsibility for addressing possible Russian subversive activities. Baltic state officials also highlight the possible link between conventional military capabilities and hostile measures, noting that a strong conventional deterrent in the region might also deter any possible lower-intensity aggression.[141] It is also likely that forces or programs that are designed to address the potential for a high-end conventional contingency might provide support for hostile measures, including intelligence or reconnaissance assets, cyber operations, and special operations capabilities. By improving situational awareness on

---

[139] U.S. Army Europe, "US Army Europe to Increase Presence across Eastern Europe," November 4, 2016; Cezary Stachniak, Twitter post, January 30, 2018.

[140] Inspector General, U.S. Department of Defense, "Evaluation of the European Reassurance Initiative (ERI)," August 22, 2017, p. 13; White House, "Fact Sheet: The United States and Estonia, Latvia, and Lithuania—NATO Allies and Global Partners," August 23, 2016.

[141] Discussions and correspondence with Estonian, Latvian, and Lithuanian officials, Tallinn; Riga; Cambridge, United Kingdom; and Washington D.C., July 2015, February 2016, December 2017, and March 2018.

land borders and in the maritime and air domains, ISR can especially improve indicators and warnings of a wide range of Russian activities.

There are also risks of increasing hostile measures associated with the increasing presence of U.S. and other NATO forces, however. One is that Russia might see such forces as a threat to its security. Russia's perspective of this threat is not necessarily correlated or related to the actual military threat that these forces might pose. Russian discourse on NATO deployments in Eastern Europe, for example, emphasizes that NATO has its own nonmilitary tools of achieving political change, and military forces could be deployed in support of a color revolution or other domestic unrest.[142] Such concerns might lead Russia to re-evaluate the potential benefit of undertaking aggressive actions in the region.

A second risk is that the presence of NATO forces in Estonia and Latvia might strengthen support for Russia among Russian speakers in the region, increasing their susceptibility to influence from Moscow. The Ministry of Defense in Estonia, for example, conducts biannual polls of its citizens. In the October 2017 poll, 59 percent of Russian speakers responded that they do not support the presence of NATO forces in Estonia, compared with only 8 percent of Estonian speakers.[143] Deploying NATO forces in an area that is mostly Russian-speaking could theoretically provoke opposition from local Russian speakers, which could, in turn, facilitate Russian hostile measures. Estonian and Latvian officials downplay this probability, observing that U.S. and NATO forces have gained the trust of Russian speakers and that Russian speakers are generally fond of military equipment and happily attend military demonstrations and parades.[144] It might

---

[142] See Bryan Frederick, Matthew Povlock, Stephen Watts, Miranda Priebe, and Edward Geist, *Assessing Russian Reactions to U.S. and NATO Posture Enhancements*, Santa Monica, Calif.: RAND Corporation, RR-1879-AF, 2017.

[143] Juhan Kivirähk, *Public Opinion and National Defence*, Estonian Ministry of Defence, April 2015, p. 50.

[144] RAND MSW symposium, Cambridge, United Kingdom, February 2, 2016; discussions with Estonian and Latvian officials, Tallinn, Riga, and Washington DC, July 2015, November 2015, January 2017, and December 2017.

be true that there would be limited risk from local Russian speakers, but this issue should be studied further.

These observations point to a few areas where the United States could engage further to improve defense and deterrence against hostile measures while limiting the potential for escalation. One such area is working with the Baltics to strengthen existing institutions. Bilateral engagement on cyber, information operations, intelligence, and interagency coordination could be useful, but the Baltics are already quite capable in many areas. The United States could also support the NATO Centers of Excellence in Tallinn, Riga, and Vilnius, focusing on cybersecurity, strategic communications, and energy security, and the European Centre of Excellence for Countering Hybrid Threats in Helsinki. While these centers are still being developed, contributions of resources and skilled personnel could go a long way toward advancing research on these topics. Perhaps the area where U.S. assistance can most directly improve capabilities is where there is an overlap between national capabilities for high-intensity military operations and hostile measures, especially because of the greater funding for U.S. Department of Defense (DoD) security cooperation and the requirement for many U.S. security cooperation programs to support U.S. military operations. Examples of overlapping capabilities include border security; internal and reserve security forces; and ISR capabilities to improve land, air, and maritime awareness. These engagements could further bolster the resilience of the Baltic states to a wide range of Russian aggression.

CHAPTER THREE

Hostile Measures in Southeastern Europe

Southeastern Europe—including the Western Balkans, Bulgaria, Romania, Moldova, and Greece[1]—is a significant target of Russian influence. U.S. interests are at stake in the region, both because some countries are NATO members (including Bulgaria and Romania), and because of continuing U.S. support for non-NATO members (including the Western Balkan countries, Ukraine, and—to a lesser extent—Moldova). Despite all this, the area has attracted less attention than Northeastern Europe (to the frustration of many pro-Western officials from the region), although senior Western officials do appear to be more mindful of the region than in the past.[2]

Russian means of influence vary extensively across the different countries in the region. In countries with large Slavic populations, for example, there are significant pro-Russian attitudes. The history of the Soviet Union also left Russia with economic, political, and social ties throughout the region that it can leverage for profit and influence. Economic difficulties, social discontent, and ethnic conflict offer other ways for Russia to exploit divisions even where there is no specific support for Russia's agenda.

---

[1] For this chapter, we include the Balkan peninsula—including Albania, Bulgaria, Greece, Moldova, Romania, and the former Yugoslavian countries—in Southeastern Europe. Turkey, although a NATO member, was excluded because it fits at the crossroads between Europe and the Middle East and is therefore subject to its own unique set of dynamics. (It is also the subject of ongoing RAND studies).

[2] Dimitar Bechev, *Rival Power: Russia's Influence in Southeast Europe*, New Haven, Conn.: Yale University Press, 2017, pp. 1–3.

Although these opportunities are significant, they do not imply that Russia can easily achieve its foreign policy objectives in the region. Although Russia's influence in EU and NATO member states—including Bulgaria, Greece, and Romania—is significant in some cases and could be used to exacerbate preexisting fissures inside NATO and the EU, it is probably not sufficient to shift the policies of these countries. Still, addressing Russian influence might be important for maintaining political stability and economic development. Russian influence is greater in non-NATO countries, such as Moldova and Serbia. Russia can continue to provoke conflict and instability in these societies, maintain its control, and undermine further EU and NATO enlargement.

Motives

Russia might have a variety of motives for pursuing hostile measures in Southeastern Europe. The five strategic interests outlined in Chapter One, including Russia's own regime security and stopping further Western enlargement, might give Moscow reason to pursue hostile measures in the region, but the economic motivations of Russian elites and companies also appear to play a key role. Although the economic vitality of Russia's elites and businesses is clearly connected to regime survival and Russia's security, these economic motivations are somewhat distinct from Russia's strategic priorities.

Perception of Threat

Russia's concerns about the region might also be motivated by the perception of a military threat from NATO and the EU. Russia has particularly emphasized that deployment of ballistic missile defense (BMD) assets in Romania, among other NATO deployments, is threatening to Russian interests and security. Putin effectively threatened Romania in May 2016 following the installation of the BMD site, noting, "If, yesterday, people simply did not know what it means to be in the crosshairs in those areas of Romania, then today we will be forced to carry out certain measures to ensure our security."[3] U.S. officials emphasize that the missiles cannot undermine Russia's deterrent, and technical analysis indicates that the sites have little utility for defending the United States against a Russian nuclear attack, but Russian officials remain unconvinced.[4] Putin explained, "At the moment, the interceptor missiles installed have a range of 500 kilometers [km], soon this will go up to 1,000 km, and worse than that, they can be rearmed with 2,400-km-range offensive missiles even today, and it can be done by simply switching the software, so that even the Romanians themselves won't know."[5] Russia's Foreign Ministry similarly warned Romania that it was "being gradually turned into yet another U.S. and NATO base near Russian borders."[6] While noting the increased presence of headquarters staff in Romania and Poland, commentary in the news outlet RT (formerly Russia Today) in late 2017 highlighted the perceived threat of the BMD sites as being greater than that of the increased rotation presence of U.S. forces in Romania.[7]

Russia's concern about missile defense plays into its broader concerns about NATO and EU enlargement in the region. As already discussed, Russia remains deeply concerned about the potential for color revolutions and the accession to EU and NATO of countries within its sphere of influence, fearing that these events could facilitate regime change in Russia and bring NATO to its doorstep.

---

[3] "Putin: Romania 'in Crosshairs' After Opening NATO Missile Defense Base," RT, May 27, 2016.

[4] See Dean Wilkening "Does Missile Defence in Europe Threaten Russia?" *Survival*, Vol. 54, No. 1, 2012.

[5] "Putin: Romania 'in Crosshairs' After Opening NATO Missile Defense Base," 2016. See also Igor Ivanov, "The Missile-Defense Mistake: Undermining Strategic Stability and the ABM Treaty" *Foreign Affairs*, Vol. 79, No. 5, September–October 2000.

[6] Quoted in Torie Rose DeGhett, "Romania Is Starting to Freak Out About Russian Designs on Transnistria," Vice News, October 6, 2015.

[7] "States Hosting Expanded NATO Forces Reduce Own Level of Security—Top Russian Diplomat," RT, December 30, 2017.

Undermining EU and NATO Enlargement

Related to its perception of threat as discussed in Chapter One, Russia seeks to stop further EU and NATO enlargement and to undermine the agendas pursued by these organizations. Southeastern Europe is an important front for EU and NATO expansion—Montenegro, for example, is close to becoming a full NATO member while Serbia, Bosnia and Herzegovina, and Moldova are all pursuing closer integration with the EU, albeit with varying success. Russia might seek to hold the line against further enlargement and prevent the frontier of NATO and the EU from moving closer to its borders or undermining Russia's sphere of influence in such countries as Moldova. Continued enlargement could set a precedent for further degradation of Russia's influence within its near abroad.

Similarly, Russia might seek to undermine the EU and NATO by targeting countries in the region that are already member states. One analyst hypothesized that Russian influence in existing EU and NATO members might be used to send the West a message that efforts to meddle in Russia's "back yard," including Ukraine, would lead to Russian activities inside the EU and NATO.[8]

Imperial Mentality

Russia's imperial mentality appears to extend to some countries within Southeastern Europe, especially Moldova, Serbia, and Bulgaria, although Russia's relationships with and perceptions of these countries vary.[9] For example, John Dunlop emphasizes that the radical Eurasianist Dugin sees the Balkan peninsula as part of Russia's natural empire and

> assigns "the north of the Balkan peninsula from Serbia to Bulgaria" to what he terms the "Russian South." "Serbia is Russia," a subheading in the book declares unambiguously. In Dugin's opinion, all of the states of the "Orthodox collectivist East" with time will seek to establish binding ties to "Moscow the Third

---

8 Discussion with Bulgarian analyst, Sofia, June 2016.

9 Discussions with Bulgarian and Romanian analysts, Bucharest, June 2016.

> Rome," thus rejecting the snares of the "rational-individualistic West." The states of Romania, Macedonia, "Serbian Bosnia," and even NATO-member Greece in time, Dugin predicts, will become constituent parts of the Eurasian-Russian Empire.[10]

Dugin's views represent an extreme view that likely does not reflect mainstream discourse, although it might shape elite views. Following the metaphor of spheres of decreasing desire for influence across Europe, Russia appears to have a stronger interest in countries in the former Soviet Union—such as Moldova, which is specifically mentioned in Russia's foreign policy concept. In the cases of Serbia and Bulgaria, Russia has stronger cultural links, given its historical and religious ties, as will be discussed later. These cultural affiliations might facilitate political or economic influence or be otherwise connected.

Economic Interests

A fundamental part of Russia's engagement in Southeastern Europe is the economic motivation of the regime and connected individuals and companies. Russian companies have major investments in the region, especially in Bulgaria, Moldova, Serbia, and Montenegro. One former Bulgarian official, for example, claimed that $5–6 billion per year came from Russia's involvement in Bulgaria.[11] This figure, amounting to approximately 10 percent of Bulgaria's GDP, might be an exaggeration—the real number is likely impossible to calculate because Russian influence over businesses and the amount of money involved are both uncertain. Whatever the figure, Russian foreign policy is likely motivated to ensure that energy companies with links to the government (including Lukoil, Gazprom, and others) continue to benefit from their involvement in Bulgaria.[12] In Serbia, Gazprom has similarly added to its dominant position in natural gas by taking a significant share of the

---

[10] Dunlop, 2004.

[11] Discussion with former Bulgarian official, Sofia, June 2016.

[12] Dimitar Bechev, *Russia's Influence in Bulgaria*, Brussels: New Direction, The Foundation for European Reform, February 24, 2016, pp. 5, 13–20.

LNG market by acquiring Serbian energy companies.[13] Russian capital and banking are also prominent in the region—in 2013, for example, foreign direct investment held by Russian entities in Southeastern Europe exceeded 5 percent of GDP in Bosnia, Bulgaria, Moldova, and Montenegro.[14]

Opportunities

Russia's opportunity for influence in the region varies widely depending on geography and the degree of shared historical, political, or economic ties.

Historical and Cultural Ties

Many countries in Southeastern Europe share a history or cultural factors that lead them to be more vulnerable to Russian influence. In some cases, the ties with Russia lead to a pro-Russian faction of society; in others, there are shared institutions or beliefs through which Russia can exert influence even if there is little pro-Russian attitude.

Two strong ties between Russia and the region are language and Orthodox Christianity. Bulgarian and Bosnian-Croatian-Serbian are South Slavic languages that are relatively closely related to Russian, which facilitates mutual communication and points to a shared history of the development of Slavic countries.[15] The Cyrillic alphabet, for example, was developed by monks in Bulgaria and later spread to other Slavic areas. Orthodox Christianity is widely practiced in Southeastern

---

[13] Marta Szpala, "Russia in Serbia—Soft Power and Hard Interests," *OSW*, October 29, 2014.

[14] Russian foreign direct investment in Montenegro is especially significant—more than 25 percent in 2013. See Larrabee et al., 2017.

[15] Croatian and Bosnian are mutually intelligible with Serbian, although they generally do not use the Cyrillic alphabet. Apart from the Bosnian Serbs, these countries are also less closely connected to Russia.

Europe, although each of the national churches are relatively independent because of the religion's decentralized character.[16]

These shared cultural ties are also closely connected to the region's history. In the 18th and 19th centuries, Imperial Russia became the supporter of Slavic and Orthodox populations inside the Ottoman Empire. Indeed, Russia's role in the Russo-Turkish war of 1877–1878 led to the independence of Bulgaria, Serbia, Montenegro, and Romania. Russia continues to use its role in helping these countries gain independence to maintain support in the region.[17] Russia, as part of the Soviet Union, also played a key role in defeating the Nazis in 1945. This shared history of World War II similarly offers Russian leadership an opportunity. In October 2014, for example, Putin visited Serbia and thanked "our Serbian friends for the way they treasure the memory of the Soviet soldiers" who fought in World War II while he also highlighted "the open manifestations of neo-Nazism that have become commonplace in Latvia and the other Baltic states."[18] In some countries, ties with Russia were reinforced by membership in Warsaw Pact, as will be discussed further.

The shared culture and history have contributed to the Slavic countries being the only countries in the region where there is a major pro-Russian faction of society and resulting domestic political support for policies friendly to Russia. In 2014, Gallup pollsters reported that 30 percent of Bulgarians supported the pro-EU forces in Ukraine; 27 percent supported the pro-Russian ones.[19] A separate poll reported that 40 percent of Bulgarians supported EU membership, and 22 per-

---

[16] The extent of the Orthodox Church's influence might be greater in Bulgaria and Serbia, where the church is more closely aligned with the practices in Russia. Bechev, 2016, pp. 3, 10; Clive Leviev-Sawyer, "Pan-Orthodox Council: Bulgarian, Russian Orthodox Church Positions 'Overlap,'" *Sofia Globe*, June 10, 2016; Kirill Ozimko, "Is Russia Losing Its Little Brother? Information War Drags Serbia Closer to EU," Fort Russ News, December 20, 2015.

[17] Ozimko, 2015; Alexander Andreev, "Bulgaria: Caught Between Moscow and Brussels," *Deutsche Welle*, April 27, 2014.

[18] Vladimir Putin, "Interview to *Politika* Newspaper," October 15, 2014.

[19] Andreev, 2014.

cent supported joining the Eurasian Customs Union.[20] Similarly, in Serbia, a former ambassador to Moscow estimated that there are "at least one-third and probably more who are always favorably inclined toward Russia and expect great things from it."[21]

Countries that have substantial numbers of Orthodox practitioners who speak a Romance language rather than Slavic one, such as Romania, tend to be less favorably inclined toward Russia but offer an opportunity for Russian influence nevertheless. Romanian analysts observed, for example, that the Russian Orthodox Church (which has links to the Russian regime) was closely connected to the Moldovan Orthodox Church and alleged that Russia was linked to anti-fracking protests organized by the church.[22] Although there was no significant pro-Russian faction in Romania, Russia could identify issues that were shared among individuals and groups, such as conservative or religious ideology, and exploit this commonality to gain support. Similarly, Putin visited Greece in May 2016 and emphasized the potential for both economic cooperation and shared cultural links through a visit to a monastery.[23]

Economic Conditions

Understanding Russia's ability to manipulate this region from an economic perspective starts with a basic fact—Southeastern Europe is poor; the poorest region of Europe. Ranking the 28 members of the EU plus the five candidate countries and one potential candidate (Bosnia and Herzegovina) by GDP per capita, Bosnia and Herzegovina

---

[20] Joe Parkinson, "Bulgaria's Western Allies Worry About Eastern Tilt," *Wall Street Journal*, May 30, 2014.

[21] Gordana Knezevic, "Wanting the Best of Both Worlds: How Serbs View Russia and EU," Radio Free Europe/Radio Liberty, May 13, 2016.

[22] Discussions with Romanian analysts, Bucharest, July 2016.

[23] Nektaria Stamouli, "Russian President Vladimir Putin Aims to Renew Ties During Visit to Greece," *Wall Street Journal*, May 27, 2016.

would rank 34th; Albania, 33rd; Serbia, 32nd; Macedonia, 31st; Montenegro, 30th; Bulgaria, 29th; Romania, 27th; and Croatia, 26th.[24]

From Russia's standpoint, this creates several direct and indirect conditions that it can exploit through hostile measures. Most directly, poverty causes discontent that Russia can attempt to turn into hostility toward Western Europe, the EU, and the West more broadly. For example, Montenegro's Prime Minister Milo Dukanovic recently made this argument about his country in the NATO Warsaw summit in July 2016. Noting the regional average per capita GDP was a mere €6,000 (euros), Dukanovic argued that Russia—through active information operations—is trying to use economic grievances in his country to fan anti-NATO and anti-EU sentiment.[25]

Poverty also causes indirect points of vulnerability to Russian hostile measures, however. Because poor economic conditions are combined with a weak rule of law, corruption and organized crime remain challenges throughout the region, giving Russia what many analysts fear is another inroad into these societies.[26] There are still other indirect ways that poverty can enable Russian hostile measures. One analyst in Bulgaria noted that low wages mean that most families rely on dual incomes, leaving child care to grandparents. Because the older generation grew up under the Soviet period and tend to be more pro-Russian as a demographic cohort, this analyst was concerned that Bulgaria's economic circumstances might inadvertently push the next generation also to adopt pro-Russian sentiments.[27]

A second set of challenges stem not from the lack of wealth but from how these countries emerged from their former Soviet-bloc selves to structure their economies. In some cases, the rocky shift to a market

---

[24] Eurostat, "GDP Per Capita, Consumption Per Capita and Price Level Indices," *Eurostat Statistics Explained*, June 2016a.

[25] Andrew Rettman, "EU Warned of Russian 'Peril' in Western Balkans," *EU Observer*, July 12, 2016.

[26] Interview with political analysts in Romania, Bucharest, June 23, 2016.

[27] Interview with political analysts in Bulgaria, Sofia, June 20, 2016; interview with Bulgarian think tank analysts, Sofia, June 21, 2016; interview with a former Romanian government official, Bucharest, June 23, 2016.

economy allowed Russian companies to buy up key infrastructure assets relatively cheaply—particularly in the energy sector.[28] In Serbia, for example, Lukoil owns 79.5 percent of Serbia's Beopetrol service-station chain and Gazprom owns 51 percent of Serbian oil reserves.[29] In Bulgaria, similarly, Lukoil acquired the oil refinery at Burgas, which is the largest refinery in the Balkans, and contributes a substantial portion of the government's tax revenues.[30] When Russian companies did not directly buy assets themselves, local proxies sympathetic to Russia often did. After the collapse of the Soviet empire, wealth was often retained among a handful of individuals, sometimes with ties back to the security services and to Russia. Weak rule of law and corruption, in turn, perpetuated these oligarchs' hold on power.[31] Ultimately, the arrangement leaves Russia in a strong position to wield economic influence.

Finally, there are structural issues that make some countries and industries in Southeastern Europe dependent on the Russian economy. The EU far outweighs Russia as a trading partner, and officials in the region emphasize that Europe is more politically and economically important than Russia. Before European sanctions, exports to Russia composed between 1 percent and 5 percent of Serbian, Slovenian, and Bulgarian GDP and more than that for Moldova.[32] Furthermore, the region is geographically close to Russia—as one senior Bulgarian government official remarked, "Bulgaria shares a common border with Russia (through the Black Sea) and a common economic interest."[33]

---

[28] Christopher Deliso, "Analysis: LUKoil in the Balkans," UPI, June 26, 2002.

[29] Nikolaus Blome, Susanne Koelbl, Peter Müller, Ralf Neukirch, Matthias Schepp, and Gerald Traufetter, "Putin's Reach: Merkel Concerned About Russian Influence in the Balkans," Spiegel Online International, November 17, 2014; Stephen Blank, "Russia's Newest Balkan Games," *Eurasia Daily Monitor*, Vol. 13, No. 47, March 9, 2016.

[30] Deliso, 2002; Center for the Study of Democracy, *Energy Sector Governance and Energy (In)security in Bulgaria*, Sofia, 2014, p. 63.

[31] Interview with Bulgarian think tank analysts, Sofia, June 21, 2016.

[32] Aasim M. Husain, Anna Ilyina, and Li Zeng, "Europe's Russian Connections," *VOX* (Centre for Economic Policy Research's Portal), August 29, 2014.

[33] Interview with a senior Bulgarian defense official, Bucharest, June 21, 2016.

As is the case in much of the rest of Europe, Russia has considerable influence as a major provider of energy in the region. Except for Greece, Romania, and Croatia, Southeastern Europe turns to Gazprom for its gas and this is unlikely to change any time soon.[34] The Transadriatic Pipeline, designed to eventually bring Azerbaijani gas to market, will not come on line until 2019.[35] Unlike many central European states, however, gas is a relatively small proportion of final energy consumption: Coal, hydropower, and—in Romania and Bulgaria—nuclear power provides major alternative energy sources.[36] In Bulgaria, for example, where officials and analysts highlight their dependence on Russian gas, natural gas provided only 13 percent of final energy consumption in 2013. Russian dominance in oil is theoretically of less concern because trade in liquid fuels is less dependent on existing infrastructure, but the dominance and influence of Russian energy companies in the region mean that switching to alternative suppliers is difficult. Bulgarian analysts, for example, emphasized that prices are significantly higher in the Balkans than in Western Europe, there is difficulty in bringing the Lukoil refinery under control of the government tax system, and the revenue from the Lukoil refinery was redistributed to maintain Russian control.[37] Although energy outlets might not offer unlimited opportunities for Russia, the existing sway of Russian companies over a variety of energy resources offers an opportunity for continued Russian influence.

Russians, similarly, are an important source of tourism for the region. Thanks to the lack of a visa requirement, Russian tourists accounted for 30 percent of overnight stays in Montenegro in 2014.[38] Some 666,538 Russians visited Bulgaria in 2014, and, despite ongoing tensions with Russia, Bulgarian Minister of Tourism Nikolina

---

[34] Dimitar Bechev, *Russia in the Balkans: How Should the EU Respond?* Brussels: European Policy Centre, October 12, 2015, p. 2

[35] Bechev, 2015, p. 2.

[36] Bechev, 2015, p. 2.

[37] Discussions with Bulgarian analysts, Sofia, June 21–22, 2016. See also Center for the Study of Democracy, 2014.

[38] "In the Balkans, NATO Has Outmuscled Russia," *The Economist*, December 11, 2015.

Angelkova addressed the Russian Duma about tourism opportunities for Russians in Bulgaria—the first such presentation since the end of the Cold War.[39] Greece has made a similar play for Russian tourists in recent years, especially after neighboring Turkey downed a Russia aircraft in Syria.[40] Even Serbia has a small but increasing Russian tourism industry.[41]

Southeastern Europe's economic challenges are projected to ease somewhat over the next five years, but the opportunities for Russian hostile measures will likely remain. In May, the International Monetary Fund noted that the region continues to experience "robust growth," and estimated GDP growth for the region was predicted at somewhere between 3 percent and 4 percent for 2016.[42] Still, not all the countries of Southeastern Europe are benefiting equally: Romania, Macedonia, and Bulgaria have been doing relatively well, but others—such as Greece and Serbia—have not.[43] Even if the growth continues apace across the region, Southeastern Europe will still remain relatively poor for the short term, and the other vulnerabilities will also continue. And although some countries in Southeastern Europe—such as Romania and Croatia—have become more energy independent, others have made less progress.[44] Above all, the problems of corruption will continue in these countries. As a result, Russia will be able to exploit economic vulnerabilities in the foreseeable future.

---

[39] Bechev, 2016, p. 6; Ministry of Tourism, Bulgaria, "In the Russian Duma: Bulgaria Is the Best Place to Accept Russian Tourists in 2016," February 25, 2016.

[40] "Bulgaria, Greece in Action to Lure Russian Tourists," *Hurriyet Daily News*, March 20, 2016.

[41] Association of Tour Operators, "Serbia Saw a 25% Increase in Tourist Arrivals from Russia in 2012," March 20, 2013.

[42] International Monetary Fund, *Central, Eastern, and Southeastern Europe: How to Get Back on the Fast Track*, May 2016.

[43] See Eurostat, "Real GDP Growth Rate—Volume," July 11, 2016b.

[44] Colin Harrison, and Zuzana Princova, "A Quiet Gas Revolution in Central and Eastern Europe," *Energy Post*, October 29, 2015; Bechev, 2015, p. 2.

Ethnic Division

Because of the timing of the breakup of the Austro-Hungarian and Ottoman empires and the varied effectiveness of nation-building efforts in the region, the boundary lines of states and ethnic groups in Southeastern Europe rarely overlap. Throughout the region, there are examples of minority ethnic groups contesting their rights or political status within their home country. Ethnic discrimination and dissatisfaction has led to a range of political divisions, separatist movements, and violent conflicts that offer a range of opportunities for Russia. Ethnic elites among ethnic minorities seek outside support in their effort to mobilize the population and demand greater autonomy or other political status. The leaders of countries that are dealing with ethnic minorities seeking autonomy or independence similarly might seek support from outside countries, including Russia. In many of these societies, there is a dual opportunity for Russia—it can undermine the Western goal of building effective and integrated states within the EU and NATO, and it can bolster pro-Russian sentiment among the particular ethnic or nationalist groups its supports.

The former Yugoslavia offers particularly significant opportunities for Russia, given the legacies of ethnic conflicts and the extensive Western goals of building peace and maintaining an open-door policy to the EU and NATO. In this respect, the reports of the radicalization of increasing numbers of European Muslims will likely facilitate Russian efforts to gain support in the region.[45] In Bosnia and Herzegovina, the postwar settlement left a segregated and dysfunctional political system. The 1995 Dayton Agreement that ended the conflict established a highly decentralized political system composed of a weak central state and two ethnically divided "entities"—the Serb-dominated *Republika Srpska* (RS) and the mainly Bosnian and Croat Federation of Bosnia and Herzegovina. According to the Dayton Agreement, the RS and its representatives effectively had veto authority over the actions of the central state. Although the international community—led by the EU and the Office of the High Representative (OHR), an inter-

---

[45] Carlotta Gall, "How Kosovo Was Turned into Fertile Ground for ISIS" *New York Times*, May 21, 2016.

national body charged with implementing the peace agreement—has sought to strengthen the central state and reverse the ethnic segregation of the war, these efforts have had mixed success. Despite significant pressure from the United States and EU, Serb officials have blocked many high-profile reform efforts and undermined efforts to strengthen the central state. Russia played a role in the peace negotiations, is a member of the steering board that oversees OHR, and has close links with Bosnian Serb officials. Russia can use its backing of the Bosnian Serb agenda to undermine U.S. and EU efforts to build an ethnically integrated Bosnia within a Euro-Atlantic institution, and to gain support among Serb populations in both Bosnia and Serbia by backing the Serb nationalist cause.[46]

In Kosovo, the legacy of war offers similar opportunities for Russia, especially through Russia's close relationship with Serbia and because NATO currently leads the peacekeeping force there—increasing its significance to the alliance as whole. Russia has historic ties with Serbia, including shared Slavic language, Orthodox religion, and Russia's support for the Serbs during World Wars I and II. Some analysts argue that Serbs were generally more pro-Russian than others in the region, partly because Serbia was not fully in the Eastern Bloc and thus did not experience Russian influence and leadership in the same way.[47] In 1999, NATO bombed Serbia to prevent continued violence in Kosovo, an autonomous region of Yugoslavia with a Serb minority and important Serbian historical and religious sites. The agreement ending the war created a UN administration to oversee Kosovo but did not specify a final status for the territory. Kosovo Albanians, who made up approximately 90 percent of the territory's population, sought independence, although Serbian authorities, backed by Russia, insisted that the territory remain part of Serbia. From 1999 to 2008, Kosovo Albanians campaigned for independence with the West, and, following riots in 2004,

---

[46] See Elizabeth M. Cousens and Charles K. Cater, *Toward Peace in Bosnia: Implementing the Dayton Accords*, Boulder, Colo.: Lynne Rienner Publishers, 2001; Florian Bieber, *Post-War Bosnia: Ethnicity, Inequality and Public Sector Governance*, New York: Palgrave Macmillan, 2006; Gerard Toal and Carl T. Dahlman, *Bosnia Remade: Ethnic Cleansing and Its Reversal*, New York: Oxford University Press, 2011.

[47] Discussion with Bulgarian analyst, Sofia, June 22, 2016.

the leading Western countries decided to initiate final status negotiations. Although Serbia and Russia continually rejected any policy that would bring about Kosovo's independence, the leading UN negotiator, Martti Ahtisaari, told Serb authorities at the beginning of the negotiations that independence was the likely outcome.[48] Again, supporting the Serbs offers Russia the opportunity both to undermine the international community's agenda and to gain support within Serbia, where there is a strong nationalist consensus against Kosovo's independence.

Beyond the former Yugoslavia, Russia could exploit the Hungarian minority in Romania, which makes up about 7 percent of the population.[49] Although this group long sought increased political rights and recognition, in part with the support of Hungary, it has generally avoided pursuing separatism.[50] Another is the Turkish community in Bulgaria, which makes up approximately 8 percent of the population and has a history of being marginalized and repressed by the Bulgarian government.[51] A third important example are the Albanian populations in Serbia, Kosovo, and Macedonia—societies where Russia has positioned itself in support of the Orthodox majority.[52] There are

---

[48] For background on the conflict, see Tim Judah, *Kosovo: What Everyone Needs to Know*, Oxford, United Kingdom: Oxford University Press, 2008; Tim Judah, *Kosovo: War and Revenge*, New Haven, Conn.: Yale University Press, 2002; Henry H. Perritt, *The Road to Independence for Kosovo: A Chronicle of the Ahtisaari Plan*, Cambridge, United Kingdom: Cambridge University Press, 2009.

[49] Center for International Development and Conflict Management, "Assessment for Magyars (Hungarians) in Romania," College Park, Md.: Minorities at Risk Program, University of Maryland, undated.

[50] One ethnic Hungarian member of the European Parliament emphasized: "Our desire for autonomy is not to be confused with the separation East-Ukrainian militants are fighting for. They don't want autonomy for the region they live in, they are militating for independence." Democratic Alliance of Hungarians in Romania, "Autonomy Does Not Mean Separation," June 16, 2014. See also Mitchell A. Orenstein, Péter Krekó, and Attila Juhász, "The Hungarian Putin?" *Foreign Affairs*, February 8, 2015; discussions with analysts, London and Bucharest, February and July 2016.

[51] Central Intelligence Agency, "Bulgaria," *World Factbook*, October 6, 2016; Tatiana Vaksberg and Alexander Andreev, "Recalling the Fate of Bulgaria's Turkish Minority," *Deutsche Welle*, December 24, 2014.

[52] Discussions with Bulgarian think tank analysts, Washington and Sofia, June 2016.

various other potential ethnic conflicts that Russia could take advantage of—indeed, analysts emphasized that Russia might identify minor groups to support in order to create instability.[53]

Means

Economic Leverage and Oligarchic Control

As already mentioned, Russia holds significant economic sway in Southeastern Europe. Russian companies own significant shares of key infrastructure, oligarchs with links to Russia control vast swaths of the economy, and Russian tourists compose a significant market niche of their economies. To date, however, Russia has focused on economic goals rather than political ones in the region. As one senior Bulgarian diplomat noted, Russia views the region in imperial terms, where periphery serves to enrich the core. In concrete terms, this analyst estimated that Bulgaria provides about $5 billion to $6 billion a year to the Russian economy and Russia wants to keep it that way.[54] So, although there are examples of Russia using its influence—particularly corruption—to advance its economic interests in the region, there are fewer clear examples of Russia using economic hostile measures for political objectives.

In Bulgaria, for example, Russia used targeted corruption to expand control over the Bulgarian energy industry. Russia's Lukoil was able to purchase Bulgaria's oil refinery in Burgas for $509 million in 1999, representing just pennies on the dollar, and now owns large swaths of land in the southern portion of the country.[55] Analysts estimate that the refinery should account for approximately 20 percent of Bulgaria's tax revenues, yet a 2015 audit indicated that Lukoil managed to avoid approximately 1 billion Bulgaria leva (about $570 million) in taxation through alleged sweetheart deals with politicians allowing for lax over-

---

[53] Discussions with United Kingdom think tank analyst, London, February 2016.

[54] Interview with former Bulgarian diplomat, Sofia, June 22, 2016.

[55] Interview with analysts in Bulgaria, Sofia, June 20, 2016; Deliso, 2002.

sight of the refinery.[56] Simultaneously, by manipulating the public bidding process for large infrastructure projects, such as the South Stream pipeline, Russian-affiliated companies earn lucrative contracts while avoiding competition.[57] Indeed, according to some think tank analysts, Lukoil spent about 150–200 million leva to sustain political proxies and maintain its favorable tax status and economic position.[58] All the while, Bulgaria pays more than Switzerland for energy.[59]

A similar story plays out with Bulgaria's nuclear reactor at Belene. The reactor was supposed to be built and operated by a Russian company, Atomstroyexport, in a deal cut by the Bulgarian Socialist Party, a party that grew out of Bulgaria's old Communist party and said to have ties with Russia.[60] The deal was canceled by a later Bulgarian government, however, amid accusations of spiraling costs and Western pressure.[61] For their part, the Bulgarian Socialists—then in opposition—labeled the cancellation of the deal as a betrayal of Bulgarian national interests.[62]

Russia also allegedly stages protests and funds lobbying groups to protect its economic prospects. Bulgarian analysts note well-documented examples of Russian money flowing through excessive spending on public relations, with excess funds being funneled to Russia's allies.[63] According to some Bulgarian think tank analysts, when

---

[56] Interview with analysts in Bulgaria, Sofia, June 20, 2016; interview with Bulgarian think tank analyst, June 21, 2016; "Customs Agency Audits Lukoil Neftochim," BTA, June 1, 2015; Bechev, 2016, p. 12.

[57] Bechev, 2016, p. 19–2; interview with Bulgarian think tank analysts, Sofia, June 22, 2016.

[58] Interview with former Bulgarian diplomat, Sofia, June 22, 2016.

[59] Interview with Bulgarian think tank analyst, Sofia, June 21, 2016.

[60] Interview with Bulgarian think tank analyst, Sofia, June 21, 2016; interview with former Bulgarian diplomat, Sofia, June 22, 2016.

[61] "22% of Bulgarians Want to Join Russia's 'Eurasian Union,'" EurActiv.com, May 15, 2015.

[62] Margarita Assenova, "Bulgaria Quits Belene Nuclear Power Plant, Open Doors to South Stream," *Eurasia Daily Monitor*, Vol. 9 No. 65, April 2, 2012.

[63] Interviews with Bulgarian think tank analysts, Sofia, June 21 and June 22, 2016.

Bulgaria considered energy diversification away from Russian energy sources, Russia sponsored public protests.[64] Similarly, when Moldova considered allowing fracking, which would have jeopardized its dependence on Russian energy, the Orthodox Church helped organize protests against the endeavor. According to some Romanian analysts, these protests were organized and at least partially funded by Gazprom.[65] The analysts say Gazprom also quietly funds lobbying groups and NGOs to push for its energy interests inside Romania.[66]

Missing from these stories, however, is much evidence that Russia is exploiting its economic advantages for political gain, as opposed to economic gain—especially through actions that would have political consequences outside Southeastern Europe—although a strategy paper prepared for Putin by the Council on Foreign Relations in Moscow did argue that Russia needed to use soft power in the Balkans. "We cannot limit ourselves to investing in companies. We must spend money on infrastructure, and for the people there who see Russia as an alternative to Western power."[67] There is some evidence of Russia investing in infrastructure, with Vladimir Yakunin, a Putin ally on the EU sanctions list, heading a project worth three-quarters of a billion euros to refurbish a 350-km stretch of Serbian rail lines.[68]

Russia also might be using its economic investments as cover for intelligence-gathering purposes. Some think tank analysts say the fact that two Russian-affiliated oligarchs own much of the Bulgarian telecommunications infrastructure might allow Russia to maintain aggressive signals intelligence collection in the country.[69] Similarly, *The Economist* reported that Germany expressed concerns that a Russian-Serbian humanitarian center—associated with the Serbian

---

64 Interview with Bulgarian think tank analyst, Sofia, June 22, 2016.

65 Interview with Romanian analysts, Bucharest, June 23, 2016; Andrew Higgins, "Russian Money Suspected Behind Fracking Protests," *New York Times*, November 30, 2014b.

66 Interview with Romanian analysts, June 23, 2016.

67 Blome et al., 2014.

68 Blome et al., 2014; Judy Dempsey, "The Western Balkans Are Becoming Russia's New Playground," *Judy Dempsey's Strategic Europe*, Carnegie Europe, November 24, 2014.

69 Interview with Bulgarian think tank analysts, Sofia, June 21, 2016.

Oil Company, which, in turn, is controlled by Gazprom, staffed with Russians granted immunity by Serbia, and outfitted with firefighting aircraft and demining equipment—might in fact be a "spy base."[70] It is difficult to substantiate these claims through open sources, however, and analysts are split over just how much Russia's economic ties to the region matter to the security situation. One Bulgarian think tank analyst asked rhetorically, "What is the difference between Russians buying companies in Bulgaria and Russians buying properties in London?"[71] Others disagree. A former senior Bulgarian analyst argues that although South Stream was canceled, Bulgarian elites eventually will need to find a way to repay the Russians for their lost investment.[72] Even in this case, however, the analyst said that appeasement would be more likely to come in the form of corrupt deals favorable to Russian interests than in the form of political support for a geostrategic issue.

Moreover, it is not clear how much Russia could turn its economic position into political power in Southeastern Europe, even if it wanted to. Despite the fact that Russia has provided a significant source of revenue for Montenegro's tourism industry and other economic sectors for decades, Montenegro still voted for sanctions against Russia over Ukraine and gained "invitee" status with NATO in May 2016.[73] Similarly, despite Bulgaria's relationship with Russia (particularly its energy sector), Bulgarian politicians yielded to pressure from the EU on the South Stream pipeline based on concern that it might face financial sanctions for violating the EU's Third Energy Package.[74] Even in Serbia, there is some question about the extent of political support that Russian economic largesse actually buys, considering the overwhelming weight of Europe on the end of the scale. As Genady Sysoev, the Balkan correspondent for Russia's *Kommersant* newspaper,

---

[70] Economist Economic Intelligence Unit, "Russian Spy Centre in Nis?" *The Economist*, November 21, 2014.

[71] Interview with a Bulgarian think tank analyst, Sofia, June 21, 2016.

[72] Interview with a former Bulgarian diplomat, Sofia, June 22, 2016.

[73] NATO, "Relations with Montenegro," May 26, 2016.

[74] "EU-Moscow Row over South Stream Gas Pipeline," BBC News, June 9, 2014; interview with a Bulgarian think tank analyst, June 21, 2016.

recently observed, "Serbia cannot turn to Russia because . . . no Serbian leadership would risk losing Western investment and aid."[75] As much economic power as Russia might have in the region, prosperity in these countries remains tied to the EU. So, although these countries would prefer to hedge their bets between the two powers, the latter tends to win out when a choice is forced between pleasing one or the other.

Finally, even if Russia could turn economic power into political influence, it is not clear Moscow would want to do so. Like any sanctions, manipulation often hurts the sender, as well as the target. Although the economies of Southeastern European countries are much smaller than Russia's, Russia still wants to profit from its investments from the region and therefore might be cautious when leveraging its economic weight for political purposes.

Political Influence

Russia is widely said to maintain extensive ties to political parties in the region. Volen Siderov's Attack Party (ATAKA) in Bulgaria might provide the best example. The party, which advocates leaving NATO and breaking Bulgaria's ties to the EU, launched its reelection campaign from Russia, coopting Russian campaign slogans.[76] According to a senior Bulgarian government official, ATAKA is said to be "completely financed by Russian sources."[77] Indeed, multiple reports suggest that ATAKA has received grants from various Russian foundations, some partially funded by the Russian state.[78] In April 2015, the Bulgarian courts opened an investigation of ATAKA for some €650,000 worth of suspicious contributions.[79] ATAKA, however, is widely said to be on the decline after its leader was accused of alcoholism and assaulting a

---

[75] Aleksandar Vasovic, "With Russia as an Ally, Serbia Edges Toward NATO," Reuters, July 3, 2016.

[76] Interview with analysts in Bulgaria, Sofia, June 20, 2016; interview with a senior Bulgarian defense official, Sofia, June 21, 2016; interview with a Bulgarian think tank analyst, Sofia, June 21, 2016.

[77] Interview with analysts in Bulgaria, June 20, 2016.

[78] Interviews with a Bulgarian think tank analyst, June 22, 2016.

[79] Bechev, 2016, p. 12

police officer.[80] Some observers even say they believe it will be out of parliament altogether by next election.[81]

Russia, however, also maintains ties to other Bulgarian political parties, although none quite so overt as ATAKA. Indeed, as one Bulgarian think tank analyst remarked, Bulgarian politicians are often eager to shop their influence on Russia.[82] A former senior Bulgarian government official agreed, although he admitted these ties were difficult to prove.[83] Perhaps the most prominent of these Russophile groups is the Movement for Rights and Freedoms, which primarily represents the Turkish community. Ahmed Dogan, who supposedly was associated with the Bulgarian security services during the Cold War, founded the party after the end of that war and has maintained a pro-Russian line since then. In fact, Turkey banned Dogan from visiting after he pushed out a fellow Movement for Rights and Freedoms politician for supporting Turkey's downing of a Russian aircraft over Syria in November 2015.[84] Other analysts have expressed concerns over the nationalist Eurosceptic and Russophile Patriotic Front party and over the Bulgarian Socialist Party, a descendent of the former Soviet-backed Communist party in Bulgaria.[85] In the latter case, the party's leadership has railed against sanctions.[86] According to 2014 polling, 34 percent of its members support joining the Eurasian Union, Russia's alternative economic sphere to the EU—a percentage only four

[80] Interview with analysts in Bulgaria, Sofia, June 20, 2016.

[81] Interview with a Bulgarian think tank analyst, Sofia, June 22, 2016.

[82] Interview with a Bulgarian think tank analyst, Sofia, June 22, 2016.

[83] Interview with former Bulgarian diplomat, Sofia, June 20, 2016.

[84] "In Moscow, Bulgarian Socialist Party Bemoans Sanctions Against Russia," *Sophia Globe*, March 18, 2015; interviews with a Bulgarian think tank analyst, June 20 and 22, 2016; "Turkey Bans Bulgarian Politicians from Entering Country—Reports," *Sofia Globe*, February 11, 2016.

[85] Interview with analysts in Bulgaria, Sofia, June 20, 2016; interview with a Bulgarian defense analyst, Sofia, June 21, 2016; interview with a senior Bulgarian defense official, Sofia, June 21, 2016; interview with a Bulgarian think tank analyst, Sofia, June 22, 2016.

[86] "In Moscow, Bulgarian Socialist Party Bemoans Sanctions Against Russia," 2015.

points shy of the right-wing ATAKA's membership.[87] Unlike ATAKA, however, evidence of Russian involvement with any of these parties is sketchy and speculative.[88]

Russia also allegedly wields political influence in other ways. In 2013, Bulgarian Prime Minister Boiko Borisov resigned after large street demonstrations took place that were reportedly set off by a spike in electricity rates, frozen wages, and corruption scandals.[89] According to some Bulgarian think tank analysts, the protests were supported by Russia out of fear that Borisov would loosen its grip on the Bulgaria energy market.[90] Borisov returned as prime minister in the fall of 2014—but, according to some Bulgarian analysts, he continues to be wary that Russia might instigate another round of protests if he defies Moscow again.[91]

Bulgaria is not unique in its political parties allegedly maintaining ties to Russia; one can find comparable stories across the political spectrum throughout the region. Greece's far-right New Dawn also takes an openly pro-Russia stance.[92] Some of these connections arise from shared economic suffering: When the Greek economy imploded and Russia felt the bite of Western sanctions, Greek and Russian politi-

---

[87] Grigas, 2014.

[88] Bechev, 2016, p. 12.

[89] Matthew Brunwasser and Dan Bilefsky, "After Protests, Prime Minister Resigns," *New York Times*, February 20, 2013; Sam Cage and Tsvetelia Tsolova, "Bulgarian Government Resigns Amid Growing Protests," Reuters, February 20, 2013.

[90] Interview with Bulgarian think tank analysts, Sofia, June 21, 2016.

[91] In a surprise move in June 2016, Borisov overruled the Bulgarian president, foreign minister, and defense minister and rejected a Romanian-backed initiative for a joint Romanian-Bulgarian-Turkish flotilla in the Black Sea under a NATO flag, functionally derailing this NATO initiative. One explanation given for his behavior was that he feared Russian-backed protests similar to the ones of 2013 (interview with analysts in Bulgaria, Sofia, June 20, 2016; interview with a Bulgarian think tank analyst, Sofia, June 21, 2016). Other analysts attribute Borisov's actions to a desire to protect Bulgaria's tourism industry (which relies on Russian tourists) and energy interests (interview with a Bulgarian think tank analyst, Sofia, June 22, 2016).

[92] Mitchell A. Orenstein, "Putin's Western Allies," *Foreign Affairs*, March 26, 2014.

cians banded together to mitigate the economic fallout.[93] Other connections stem from perceived cultural ties. In May 2014, Golden Dawn members Artemis Mattheopoulos and Eleni Zaroulia tried to align the "formal approach of Hellenism with Orthodox Russia."[94] And yet, there are also allegations of more-covert Russian involvement. Michael Orenstein, for example, alleged in *Foreign Affairs* that Golden Dawn is "thought" to receive funds from Russia.[95]

Russia also allegedly developed ties to the ruling radical left Syriza party. Indeed, when Greek Prime Minister Alexis Tsipras assumed office, the first foreign leader that he invited to meet with was Russian Ambassador Andrey Maslov, and Greek Foreign Minister Nikos Kotzias chose Russia for his first non-EU trip.[96] Both Kotzias and the defense minister, Panos Kammenos (who is also the head of Syriza's coalition partner, the right-wing populist Independent Greeks

---

[93] Carol J. Williams, "Russia and Greece Consider Collaborating to Circumvent Western Sanctions," *Los Angeles Times*, June 21, 2015. Similarly, the BBC reported that "a drove of Greek cabinet members will be heading to Moscow" in hopes of courting Russian aid. Giorgos Christides, "Could Europe Lose Greece to Russia?" BBC News, March 12, 2015. Public opinion polls show a closer Greco-Russian relationship also might have become increasingly politically popular during that time. Gallup opinion polling from 2010 through at least 2014 indicated that more Greeks approved of Russia's leadership than the EU's by steadily larger margins (from 2 percent in 2010 to 12 percent in 2014) and 2013 Pew polling found that 63 percent of Greeks held a favorable view of Russia—the highest of any of the 38 countries surveyed and more than 20 points higher than the other seven European countries surveyed in the report. Phoebe Dong and Chris Rieser, "More Greeks Approve of Russia's Leadership Than EU's," Gallup, February 2, 2015; "Global Opinion of Russia Mixed: Negative Views Widespread in Mideast and Europe," Pew Research Center, September 3, 2013.

[94] David Patrikarakos, "The Greeks Are Not 'Western:' Greece and Russia Breathe New Life into Their Ancient Eastern Alliance," *Politico*, April 22, 2015; also see Antonis Klapsis, *An Unholy Alliance: The European Far Right and Putin's Russia*, Brussels: Wilfried Marten Centre for European Studies, 2015, p. 19.

[95] Orenstein, 2014. For similar allegations (not confirmed by hard evidence) of Russian funding of Golden Dawn, see Klapsis, 2015, p. 28; and Peter Foster and Matthew Holehouse, "Russia Accused of Clandestine Funding of European Parties as US Conducts Major Review of Vladimir Putin's Strategy," *The Telegraph*, January 16, 2016.

[96] Sam Jones, Kerin Hope, and Courtney Weaver, "Alarm Bells Ring over Syriza's Russian Links," *Financial Times*, January 28, 2015; "Conscious Uncoupling," *The Economist*, April 3, 2014.

party), allegedly have ties to Putin's inner circle.[97] According to emails obtained by the German newspaper *Die Zeit*, Russia's oligarchs maintain close ties with Greek politicians across the political spectrum.[98] Some correspondence shows that Syriza officials have coordinated strategy and public relations with author Dugin and Russian oligarch Konstantin Malofeyev—although there is no concrete evidence that Russia provided political support.[99] According to the *Financial Times*, the relationship between the Greek government and Russia extend beyond normal diplomatic relations, so much so that European and NATO counter-intelligence officials have voiced concerns about these ties.[100] Unsurprisingly, Greek officials have also spoken in opposition to Russian sanctions, with the energy minister, Panagiotis Lafazanis (who is also the leader of Syriza's far-left faction), stating bluntly, "We have no differences with Russia and the Russian people."[101]

Russia also maintains friendly relations with a number of Serbian political parties.[102] Two right-leaning parties—the Democratic Party of Serbia and the Dveri Movement—visited Crimea in 2015 and proclaimed that "Crimea is part of Russia just like Kosovo is a part of Serbia."[103] More recently, three Serbian parties—the Democratic Party of Serbia, the Dveri Movement, and the Serb People's Party—traveled to Russia to sign a declaration of "military neutrality" with the ruling United Russia party and expressing support for expanded Russian influence in the region.[104] Similar

---

[97] Jones, Hope, and Weaver, 2015.

[98] Meike Dülffer, Carsten Luther, and Zacharias Zacharakis, "Caught in the Web of the Russian Ideologues," *Die Zeit*, February 7, 2015.

[99] Robert Coalson, "New Greek Government Has Deep, Long-Standing Ties with Russian 'Fascist' Dugin," Radio Free Europe/Radio Liberty, January 22, 2015.

[100] Jones, Hope, and Weaver, 2015.

[101] Jones, Hope, and Weaver, 2015.

[102] See Jelena Milić, "Putin's Orchestra," *The New Century*, No. 7, May 2015.

[103] "Serbian Delegates Say They Recognize Crimea Part of Russia," Tass Russian News Agency, October 27, 2015.

[104] Beta, "Serbian Parties Sign Declaration with United Russia," B92, June 29, 2016.

declarations have been signed by Montenegro's New Serbian Democracy, the Democratic People's Party, the Socialist People's Party, and the Bosnian Serb Alliance of Independent Social Democrats, led by Milorad Dodik.[105] Importantly, as with many of the other cases, there is no conclusive evidence of direct Russian funding of these parties—although one leader of the Democratic Party of Serbia, Nenad Popovic, runs the ABS Electro group, which has manufacturing and sales operations in Russia.[106]

Overall, Russia enjoys the support of several political parties—mostly, although not uniformly, on the far right of the political spectrum—across multiple countries in Southeastern Europe. These parties make no effort to conceal the fact that they advocate a larger Russian presence in their region, and many of these parties have already proven useful to Russia in small ways—from legitimizing its actions in Crimea to pushing against sanctions. Southeastern Europe's economic problems and immigration crisis have contributed to electoral success for many of these parties, and their support might strengthen over the next five years. That said, it can be difficult to determine the parties over which the Russian state can exercise influence, as opposed to which parties simply share similar foreign-policy orientations—although there are a handful of exceptions, such as ATAKA (whose star seems to be fading).

Information Operations

Russia plays an active role in the information space of Southeastern Europe, although its presence in that sphere might be for cultural, economic, and historical reasons rather than a strategic attempt at manipulation. This makes it difficult to evaluate the threat posed by Russia's information activities in the region. Given that a significant portion of Southeastern Europe's population understands Russian and speaks Slavic languages, segments of this smaller group—particularly in such places as Bulgaria, Moldova, and Serbia—

[105] Beta, 2016.

[106] Fairclough, Gordon, "EU and Russia Loom Over Serbian Election," *Wall Street Journal*, May 4, 2012.

naturally turn to Russia for news and culture. Although much of the content might be relatively benign, there is some evidence of Russian propaganda and more-deliberate attempts to control media content in the local language.

Russia maintains an active media presence in Southeastern Europe almost by default. In Bulgaria, for example, much of the population grew up speaking Russian and so can naturally turn to Russian outlets as a source of foreign news. There are also indirect ways for Moscow's messages to enter Southeastern Europe. For example, many Bulgarian media outlets are too poor to afford their own foreign correspondents and investigative journalists, so they often turn to Russian news outlets as sources of information for their own stories.[107] Troublingly, this practice also occurs at some prominent media outlets—perhaps most notably pan.bg, a defense and security policy site regularly read by many Bulgarian military officers that commonly features articles taken directly from Russian sources.[108] In Moldova, the story is much same. Many reporters already read Russian and some are friendly to Russia, so they turn to Russian sources for content.[109] In both Moldova and Bulgaria, the net result is that local media ends up adopting a Russian slant.

However, Russia also tries to shape the media market of Southeastern Europe directly. When Serbia's Tanjug news agency closed in November 2015, the government-funded Sputnik news service stepped in to the fill the information void in Serbia and the ethnically Serb parts of Bosnia and Kosovo.[110] Indeed, Sputnik had ramped up its Ser-

---

[107] Interview with a Bulgarian think tank analyst, June 21, 2016.

[108] Interview with a Bulgarian think tank analyst, June 22, 2016; interview with a senior Bulgarian defense official, June 21, 2016.

[109] Interview with Romanian analysts, Bucharest, June 23, 2016; Marija Šajkaš and Milka Tadić Mijović, "Caught Between the East and West: The "Media War" Intensifies in Serbia and Montenegro," Washington, D.C.: Center for International Media Assistance, National Endowment for Democracy, March 10, 2016.

[110] Andrew Rettman, "Western Balkans: EU Blindspot on Russian Propaganda," *EU Observer*, December 10, 2015.

bian broadcasts since 2014.[111] RT also recently expanded its Serbian language variant to cover the Serbian and Montenegrin markets.[112] Russian media outlets also broadcast in Bulgarian through the Voice of Russia radio and *Ruski Dnevik (Russian Diary)*, part of Russia Beyond the Headlines.[113]

There are also sporadic reports of indirect attempts by Russia to expand its control over the media. For example, the New Bulgarian Media Group owns several of the most popular newspapers and weekly periodicals and has ties to the Russian-linked Movement for Rights and Freedom party.[114] In Moldova, by some estimates, up to 85 percent of the media outlets in the country are owned by Russian-affiliated oligarchs.[115] Russians—or their affiliates—also dominate the Montenegro media market, running several major radio stations that service the entire country.[116] And in February 2016, news outlets reported that Malofeev was attempting to buy Serbian television stations through a local businessman.[117]

Perhaps, more troubling, however, are Russia's more-covert attempts to influence the media, particularly in the online space. In Romania, for example, overt Russian propaganda is less of a problem because Romanian is a Latin language and a smaller percentage of the population reads Russian.[118] Instead, analysts claim that Russia tries to influence public discourse via internet "trolls," individuals paid to rebut articles perceived as hostile to Russian interests by posting a

---

[111] Branka Mihajlovic, "Russian Seeking Serbian Media Outlet?" Radio Slobodna Europa, February 14, 2016.

[112] Mihajlovic, 2016.

[113] Bechev, 2016, p. 22.

[114] Vesela Tabakova, "Media Landscapes: Bulgaria," Maastricht, The Netherlands: European Journalism Centre, undated; Bechev, 2016; interview with analysts in Bulgaria, Sofia, June 20, 2016.

[115] Interview with Romanian analysts, Bucharest, June 23, 2016.

[116] Interview with Romanian analysts, Bucharest, June 23, 2016; Šajkaš and Mijović, 2016.

[117] Mihajlovic, 2016.

[118] Interview with Romanian Ministry of Foreign Affairs officials, Bucharest, June 23, 2016.

response in the comments section, sometimes within an hour of publication.[119] According to some Romanian media analysts, many "trolls" are freelancers rather than ideologues. They are paid approximately $500 a month for their services and some were previously paid by various Romanian political parties to perform a similar function before switching to work for Russian interests.[120] One Romanian media analyst said there are at least 20 such groups of trolls operating in the country.[121] Analysts say that these "trolls" support Russia's goals of undermining pro-Western views and complement Russia's efforts to strengthen special interest groups that agree with Russia on particular issues. Romania is not alone in this regard: Some Bulgarian think tank analysts say the most important Russian information operations today are distributed online and via social media rather than through more-traditional news outlets.[122]

Through these overt and covert media links, Russia spreads a variety of messages. According to Russia analyst Stephen Blank, "The Russians are exploiting a flood of ethno-religious propaganda stressing the supposed unity of the Serb and Russian peoples (though this is far less than the Kremlin makes it out to be), the threat from NATO and the European Union, the lure of Russian energy, and unresolved issues in Kosovo."[123] Other analysts argue that Russia planted stories in Serbia to suggest that the shooting down of a Malaysian airliner by Russian separatists was, in fact, the work of the Romanian military and to undermine support for the sanctions against Russia.[124] In Bulgaria, Russia used its media ties to bolster support for its Syria intervention, undermine fracking (which would come at the expense of Russian energy imports) and boost its efforts to build the South Stream pipe-

[119] Interview with Romanian analysts, Bucharest, June 23, 2016.

[120] Interview with Romanian analysts, Bucharest, June 23, 2016.

[121] Interview with Romanian analysts, Bucharest, June 23, 2016.

[122] Interview with Bulgarian analyst, Sofia, June 21, 2016.

[123] Stephen Blank, "Putin Sets His Eyes on the Balkans," *Newsweek*, April 17, 2015.

[124] Mihajlovic, 2016.

line.[125] In Romania, government officials assess that Russia has tried to discredit EU integration efforts by planting stories on poor living conditions in Bucharest.[126]

Russia also maintains an active cultural diplomacy effort in the region. In 2013, Russia opened a cultural center in Belgrade run by the Russian Foreign Cooperation Agency, and several other nominally independent Russian organizations operate in Serbia, including the Gorchakov Public Diplomacy Fund, the Strategic Culture Foundation, the Centre of National Glory, the Foundation of St. Andrew, and the Fund for the Russian Necropolis.[127] Even in non-Slavic Romania, Russia funds a cultural center in Bucharest that regularly holds events highlighting Russian culture.[128] Although these efforts could be dismissed as an attempt to build public goodwill similar to embassies elsewhere the world over, these cultural efforts take on a more political tinge. In Bulgaria, for example, the English version of the right-wing Russian-backed ATAKA's website features a press release detailing how it closed the 2014 election cycle with a concert in which "Putin's favourite musicians came to Bulgaria to perform their world-famous hits."[129] The evening concluded by Yosif Kobzon, a Russian music star turned Duma member, who thanked ATAKA for its position on the Ukraine crisis and reminded the audience, "Russia liberated Bulgaria from Ottoman Yoke, not the European Union."[130]

Russia also has inroads into academic institutions. According to some Romanian think tank analysts, Russia maintains at least indirect ties to left-wing academics (particularly in Kluge), stoking pacifist and

[125] Interview with Bulgarian analyst, Sofia, June 22, 2016; interview with former Bulgarian diplomat, Bucharest, June 21, 2016.

[126] Interview with Romanian officials, Bucharest, June 24, 2016.

[127] Szpala, 2014.

[128] Interview with Romanian analysts, Bucharest, June 23, 2016.

[129] "Kobzon Thanked Siderov for his Position on Behalf of Russia and Wished ATAKA Success at the Elections," ATAKA, May 22, 2014.

[130] "Kobzon Thanked Siderov for his Position on Behalf of Russia and Wished ATAKA Success at the Elections," 2014.

anti-U.S. sentiments rather than pro-Russian ones.[131] Elsewhere, the ties might be somewhat more nefarious. Bulgarian think tank analysts suggest that Russia maintains ties with the "Library Studies" department of a major Bulgarian university as cover for recruiting and training schools for the Bulgarian security services.[132]

The question, of course, is what Moscow gains from its information operations. Some polls suggest that the effort might be yielding moderate success. Support for EU integration in Moldova has declined to slightly more than half the population, down from 72 percent seven years ago.[133] According to 2014 polling, Serbs view Russia more favorably than any other foreign power (52 percent to 17 percent); according to a 2016 study by the Belgrade-based Center for Euro-Atlantic Studies, 42 percent of young Serbs would like to implement a Russian political system in Serbia (ironically, 70 percent would also live elsewhere in Europe or the United States if given the choice).[134] Public opinion polling in Bulgaria shows a similarly mixed result. A 2015 poll showed that 61 percent of Bulgarians view Russia equally or more favorably than they did before the Ukraine crisis, significantly more than the roughly 27 percent of Europeans overall.[135] At the same time, 63 percent of Bulgarians would choose NATO and the EU over Russia and the Eurasian Union in a hypothetical matchup.[136] Ultimately, a range of cultural, historic, and economic factors could influence these polls, so it is hard to isolate the effect of Russia's information operations on public opinion. Still, at a surface level, it seems plausible that Russian outreach efforts have helped at least sustain pro-Russian sentiments in the region.

---

[131] Interview with Romanian analysts, Bucharest, June 23, 2016.

[132] Interview with Bulgarian analysts, Sofia, June 22, 2016.

[133] Interview with Romanian analysts, Bucharest, June 23, 2016.

[134] Bechev, 2015, p. 3; Knezevic, 2016.

[135] European Council on Foreign Relations, "Public Opinion Poll: Bulgarian Foreign Policy, the Russia-Ukraine Conflict and National Security," March 26, 2015.

[136] European Council on Foreign Relations, 2015.

Encouraging Ethnic Conflict and Separatist Movements

Russia can support ethnic or separatist movements or offer support for the majority groups that oppose these movements, to achieve its goals in the region.

In the case of Russian support for a separatist movement or an ethnic group seeking autonomy, Russian behavior generally follows a pattern resembling that of conflicts outside the region, including in the Donbas in Ukraine and Abkhazia or South Ossetia in Georgia. In these conflicts, Russia seeks to create autonomous regions, frozen conflicts, or complex federal arrangements to make it difficult for the host country to exercise effective governance or to achieve the requirements of the EU and NATO for accession. For example, NATO specifically requires that countries aspiring to membership resolve existing "ethnic disputes, external territorial disputes, including irredentist claims or internal jurisdictional disputes, by peaceful means,"[137] thus implying that countries such as Moldova, Georgia, and Ukraine will have to resolve the existing separatist movements prior to gaining membership. If Russia can give sufficient support to separatist movements to prevent these countries from resolving the disputes, it can effectively block NATO accession. Furthermore, if an autonomous region gains the right to veto some decisions at the national level, that region can undermine the functioning of the government, thereby undermining the progress of Euro-Atlantic integration, as in Bosnia. Finally, even when ethnic or separatist movements cannot achieve specific political rights or authorities, they can undermine the efforts of host countries to build a domestic consensus and unified national identity.

Alternatively, Russia might support the leadership of a country, often a majority ethnic group, in opposition to an ethnic or separatist movement. In these cases, Russia might have multiple goals or motivations, such as gaining the support of the majority community, working with Slavic or Orthodox allies, and avoiding encouragement of a precedent of separatism to limit the prospects of separatism in Russia. Russian decisions to intervene in a conflict, and the resources put forward to do so, are necessarily ad hoc calculations based on its interests

[137] NATO, "Study on NATO Enlargement," September 3, 1995.

and available resources, so it should not be surprising that Russia has supported separatist movements at some times and opposed them at others.

In Southeastern Europe, there are three major ethnic and separatist conflicts in which Russia has intervened and likely could again in the future: Bosnia and Herzegovina, Kosovo, and Moldova. Analysts in the region note that "Russia can take its pick" of intervention in these high-vulnerability cases. Based on its military presence or status as a permanent member of the Security Council, Russia can block a solution in each of these conflicts or escalate a conflict when it so chooses to distract Western attention and discredit NATO and EU peacekeeping missions in Bosnia and Herzegovina and in Kosovo.[138]

In Moldova, Russian has supported the separatist region of Transnistria since 1992, and the region still serves as a so-called control button for Russia to manipulate the situation in the region.[139] Russia has prevented the integration of Transnistria into Moldova through the presence of troops in the region, economic support for the Transnistrian government, and provision of Russian passports to Transnistrians. Russia has also begun to support another potential separatist region of Moldova: the Autonomous Region of Gagauzia. Although Moldova is not a major priority for the United States or NATO, Romania has close connections with Moldova and will continue to engage there and possibly be drawn into a military role if the situation in the country declines.[140]

In Bosnia, Russia will likely continue to undermine political progress through its support for the Bosnian Serbs and could increase this support to destabilize conditions there. Russia has several means of action in the region. First, Russia can use its position on the Steering Board of the Peace Implementation Council, the body that oversees the peace settlement and continues to supervise the OHR, to undermine the Dayton Agreement and ethnic integration of Bosnia. Second, Russia could use the UN Security Council as a vehicle to undermine

---

[138] Discussion with Romanian officials, Bucharest, June 2016.

[139] Discussion with Romanian officials, Bucharest, June 2016.

[140] Discussion with Romanian officials, Bucharest, June 23–24, 2016.

Western policy in Bosnia. In July 2015, Russia vetoed a UN Security Council resolution that would have labeled the 1995 Srebrenica massacre as a genocide and abstained from a vote in November 2014 on extending the mandate on the EU peacekeeping mission in Bosnia.[141] Third, Russia has links with the leadership of the Serb-led RS and has supported RS President Dodik in blocking Western efforts to strengthen the central state in Bosnia. For example, in November 2015, Russia supported Dodik's plans to hold a referendum in February 2016 on the authority of OHR to impose and implement laws. The referendum was indefinitely delayed following international pressure. Nevertheless, Dodik, with Russian support, continues to be able to undermine the Western agenda in Bosnia and could resurrect the same referendum or implement other means of escalating the conflict to destabilize Bosnia.[142]

James Lyon, who formerly oversaw the International Crisis Group's efforts in the Balkans, explains one view of risks of Russian involvement in Bosnia, noting that "by backing Dodik, Putin is able to create substantial problems for the West without needing to invest resources or diplomatic energy." Lyon warns that "the West must now prevent Russia from using Dodik's nationalist agenda to destabilize the Balkans and create yet another proxy conflict."[143] In January 2018,

[141] Somini Sengupta, "Russia Vetoes U.N. Resolution Calling Srebrenica Massacre 'Crime of Genocide,'" *New York Times*, July 8, 2015; Tanjug, "Russia Abstains During Vote to Extend EUFOR Mandate," B92, November 12, 2014; Elvira M. Jukic, "Russia Flexes Muscles on EU Bosnia Mission," *Balkan Insight*, November 17, 2014.

[142] Danijel Kovacevic, "Bosnian Serb Leader Postpones Controversial Referendum," *Balkan Insight*, February 9, 2016.

[143] James Lyon explains,

> The Russian ambassador to Bosnia, Pyotr Ivantsov, has stated that the referendum is an internal matter for the country and has expressed his sympathy toward Republika Srpska complaints over the state judiciary. Russian ambassadors have been notable in their refusal to support the international community's efforts to stop Dodik's attempts to tear Bosnia apart, as well as in their opposition to Bosnia's EU and NATO membership, issues that they had earlier agreed to. . . . There is now even some question as to whether Russia supports the sovereignty and territorial integrity of Bosnia.

James Lyon, "Is War About to Break Out in the Balkans?" *Foreign Policy*, October 26, 2015.

members of the Serbian Honor group (who were alleged without specific evidence to be a Russian-trained paramilitary group "organized to act against Dodik's political opponents") marched in a parade in Banja Luka, the RS capital.[144] Still, it is important not to overestimate the immediate threat posed by Dodik and the Bosnian Serbs. Dodik had repeatedly threatened secession and supported referenda over the past few years and been pulled back, partly because the Bosnian Serbs, like others in Southeastern Europe, cannot afford to alienate the EU as a result of Bosnia's economic dependence on Europe and the consensus view within the country that its future lies within the EU.

Russia has a similar range of tools it can use to achieve its goals with regard to Kosovo. As discussed, Russia's primary interests lie in supporting Serbia's efforts to prevent Kosovo from gaining independence. Russia's policy in Kosovo is likely guided by at least four goals: undermining EU and NATO efforts to build peace and integrate Kosovo, working in support of a Slavic and Orthodox country for ideological reasons, opposing the precedent of separatism in order to limit separatism within Russia, and seeking the support of Serbia and other Serbs in the region for Russia's political interests in Europe.[145] Russian officials clearly have an incentive to dissemble their motivations, but Russian action is likely guided to some degree by each of these factors.

As in Bosnia, the first means that Russia can use to undermine progress in Kosovo is Moscow's official position on international bodies. Russia used its veto on the UN Security Council in 2008 to prevent official recognition of Kosovo's independence and to prevent the UN from transforming the UN Mission in Kosovo to reflect Kosovo's declaration of independence.[146] Russia's obstruction facilitated further European divisions about Kosovo, with the result that the nation remains in a legal

---

[144] Maja Zuvela, "Bosnia to Investigate Suspected Serb Paramilitary Group," Reuters, January 16, 2018.

[145] See Judah, 2008, pp. 135–139; Oksana Antonenko, "Russia and the Deadlock over Kosovo," *Survival*, Vol. 49, No. 3, 2007.

[146] International Crisis Group, *Kosovo's Fragile Transition*, Europe Report, No. 196, September 25, 2008, p. 3.

gray area.[147] Indeed, five members of the EU have not recognized Kosovo's independence;[148] compounded with bureaucratic challenges, this has undermined the effectiveness of the EU's efforts there.[149] The EU has used Serbia's membership aspirations as an incentive to encourage dialogue between Serbia and Kosovo,[150] which implies that Russia could continue to use its influence in Serbia and the region to undermine EU efforts at integration. Although Kosovo is de facto independent, the UN Mission in Kosovo theoretically still has sovereign authority and official documents referring to Kosovo must note that UN Resolution 1244 (the 1999 resolution that ended the war in the territory) is still in effect.

Second, Russia can offer economic or military assistance to Serbia and the Kosovo Serbs. Although offering some support, Russia has not to date invested major resources in coming to the aid of the Kosovo Serbs. In 2011, for example, Serbs in northern Kosovo petitioned for Russian citizenship; their request was declined but Russia did send a convoy with 284 tons of humanitarian aid.[151] Similarly, analysts note that Russia hopes Serbia will be a market for high-technology military equipment, including S-300 anti-air systems, but will seek to make a significant profit on such sales.[152]

Third, Russia can bolster its political relationship with Serbia in hope of using support for Kosovo to gain Serb backing for Russian

---

[147] One Russian commentator noted, "Russia realizes that any unilateral declaration of independence for Kosovo that does not follow UN procedure will not be recognized by all members of the European Union, and could cause a rift within the bloc." Quoted in Judah, 2008, p. 138.

[148] The nonrecognizing countries are Spain, Romania, Slovakia, Greece, and Cyprus. Spyros Economides, James Ker-Lindsay, and Dimitris Papadimitriou, "Kosovo: Four Futures," *Survival*, Vol. 52, No. 5, 2010.

[149] See, for example, Andrew Radin, "Towards the Rule of Law in Kosovo: Why EULEX Should Go," *Nationalities Papers*, Vol. 42, No. 2, March 2014.

[150] European Union External Action, "Normalisation of Relations between Belgrade and Pristina," undated.

[151] "Russia Gives Blankets to Kosovo Serbs Instead of Citizenship," *Pravda*, December 8, 2011.

[152] Discussion with Bulgarian analyst, Sofia, June 2016; "Russia's S-300 Missile Systems 'Too Costly' for Serbia—PM," Tass Russian News Agency, January 11, 2016.

activities. During his visit to Serbia in October 2014, Putin emphasized that Russia would never recognize Kosovo as part of a reaffirmation of friendship between Russia and Serbia. Serb President Tomisalv Nikolić in turn emphasized that Serbia would not join European sanctions against Russia and that "Serbia will not endanger its morality by any hostility towards Russia."[153] In March 2016, Nikolić similarly emphasized that "Putin said clearly that Serbia can count on Russia everywhere that Serbia is defending its territorial integrity and independence, especially in Kosovo."[154] Russia probably cannot change Serbia's future with the EU, but it can feed Serbian dissatisfaction regarding NATO, as it has done through Russian media.[155]

There are other, less prominent ethnic conflicts in the region that Russia also could attempt to exploit. In theory, Russia could encourage opposition by the Hungarian minority within Romania, in part through its links with Prime Minister Viktor Orbán. There also possible scenarios of Russians supporting a Hungarian breakaway region in central Romania.[156] As discussed, however, most analysts see little evidence of Hungarian minorities in Romania acting in support of radical views or being easily coopted by Russia.[157]

---

[153] A *Guardian* article noted, "An enthusiastic crowd, estimated by the Serbian government as 100,000-strong, lined the parade route and chanted "Putin, Putin," and "Serbia-Russia, we don't need the [European] Union." Julian Borger, "Vladimir Putin Moves to Strengthen Ties with Serbia at Military Parade," *The Guardian*, October 16, 2014; Putin, 2014.

[154] "Putin Promises Serbian Leader Russia Will Back Claim to Kosovo," Radio Free Europe/ Radio Liberty, March 11, 2016.

[155] For example, one article draws a direct link between the terms of a recent agreement between the Serbian government and NATO and an annex to the 1999 Rambouillet agreement that Serbia rejected, leading to the start of bombing, which permitted NATO personnel unlimited access to Serbia. Sergey Belous, "How Long Will Belgrade Seesaw Between NATO and Russia?" *Oriental Review*, April 23, 2016.

[156] Discussions with Romanian analysts, Bucharest, June 2016.

[157] Orenstein, Kreko, and Juhasz, 2015. The authors note that

> most Hungarians abroad are not supportive of aggressive autonomy movements, as they understand that they would be the first victims of irredentism. They tend to support political forces that are working towards peaceful cooperation among Hungarian, Romanian, Serbian, and Slovakian political forces.

Military Influence

Over the past several years, Russia has already bolstered its military presence in the Black Sea, boosting both its air and maritime presence in the region. Of particular concern to NATO, Russia has strengthened its anti-access and area-denial capability based out of Crimea. Long-range anti-air and anti-ship missiles will pose a major threat to NATO forces operating in and around the Black Sea and could undermine reinforcement of the area.[158] Russian military demonstrations also remain a concern throughout the region. For example, Russia has exercised its rights under the Treaty on Open Skies to fly over Romania and regularly conducts military exercises in the Black Sea.[159] These exercises might have dynamics of intimidation, as in the Baltics, and the potential for escalation in the region remains.

Although generally skeptical of the likelihood of large-scale or overt Russian military action in NATO countries in the region, Bulgarian and Romanian analysts speculate that Russia might someday engage in covert or limited military action. For example, Russia or its proxies could temporarily seize a primarily Russian beach resort or other target as a show of force, to "tease" NATO.[160] One analyst cited concerns about Russian-backed paramilitaries in Bulgaria. These groups likely draw from border patrol groups that were originally formed to combat migrants. Russia could support and incite these groups to undermine Bulgarian support for NATO and might be able to expand or reinforce this group.[161] Russian forces or proxies could

---

[158] Discussions with Bulgarian and Romanian officials, June 2016.

[159] Discussions with Romanian Ministry of Foreign Affairs officials, June 2016.

[160] Interview with Romanian analyst, Bucharest, June 23, 2016.

[161] Interview with Bulgarian analyst, Sofia, June 22, 2016. A Bulgarian news report noted,

> There also appears to be an overlap between the paramilitary patrols of the border and some of the organisations that turned out in Bulgaria last week to welcome Russia's pro-Putin "Night Wolves" motorcycle group, that resulted in confrontations and arrests in Bourgas and Sofia.

"Concern Grows over Bulgarian Paramilitaries and 'Border Patrols,'" *Sofia Globe*, July 7, 2016.

also attack energy infrastructure to undermine regional efforts at countering Russia's energy monopoly.[162]

Still, most Romanian officials highlighted that non-NATO countries—such as Serbia, Moldova, Ukraine, and Georgia—were more vulnerable to Russian military aggression and more central to Russian interests than either Romania or Bulgaria.[163] Russia could attempt to expand the conflict in Ukraine, possibly building a land bridge to Transnistria or supporting protests in Odessa.[164]

Breaking EU or NATO Consensus

A commonly hypothesized risk from Southeastern Europe is that Russia will use its influence within the region to undermine EU and NATO consensus. For example, the argument that Russia supports the accession of Serbia to the EU (and that Russia historically sought the accession of Bulgaria[165]) as a "Trojan horse" assumes that Russia will be able to undermine the activity of these organizations by exercising leverage over them.

The Trojan horse strategy is unlikely to succeed, however, because of the importance that EU and NATO have for the countries in the region. Bulgaria, Romania, and Greece are the beneficiaries of substantial amounts of EU funds, for example—the difference between EU spending and contributions in 2015 was €1.5 billion in Bulgaria, €4.1billion in Romania, and €5.3 billion in Greece. Analysts in Bulgaria emphasize that these funds are critical for the popularity and sustainability of the respective governments.[166] A given country's government or leadership might fear that unilaterally undermining the

---

[162] Interview with Romanian analysts, Bucharest, June 23, 2016.

[163] Discussions with Romanian officials, Bucharest, June 2016.

[164] Discussion with Romanian and Ukrainian officials, Kyiv and Bucharest, May 2015 and June 2016.

[165] One Bulgarian analyst noted that Russia recognized that it could not stop EU accession and sought "Bulgaria as a Trojan horse" in the EU, although this strategy did not work out in the end. Discussion with Bulgarian analyst, June 22, 2016.

[166] European Commission, "EU Revenue and Expenditures," Excel file, 2015; interviews with Bulgarian analyst and Western analyst, Sofia, June 20 and 22, 2016.

consensus of the EU would lead the EU to cut off support, undermining that country's economic future. Similarly, officials in the region emphasized that NATO is fundamental to their defense and political plans and that unilaterally vetoing a NATO resolution would undermine their relationships with NATO and future defense strategies.[167]

Thus, the risk of a Southeastern European country breaking consensus and undermining the effectiveness of the EU or NATO is quite small. As one Bulgarian analyst explained, "There is no danger to the West from Bulgaria—rather, the danger [from Russian influence] is to Bulgaria."[168] Instead of these countries presenting a vulnerable point that Russia can use to undermine the rest of the EU and NATO, the weaknesses that Russia can exploit seem far more likely to be felt locally without substantial spillover to the rest of Europe.

The Future of Russian Hostile Measures in Southeastern Europe

Overall, Russia has significantly more ability to conduct hostile measures in Southeastern Europe than in other parts of Europe, although this area—except for the non-EU and non-NATO countries in the region (including Serbia and Moldova)—is perhaps less vulnerable than Baltics. The mixture of economic, cultural, and historical connections, coupled with the fact that many countries in this region suffer from poverty and weak rule of law, provide Russia with significant opportunities. That said, Russia's ability to influence this region is not absolute. Three caveats particularly need to be kept in mind.

Vulnerability to Hostile Measures in Southeastern Europe Varies Extensively by Country

Non-EU and non-NATO countries in the region, especially those closer to Russia, are subject to a wide range of Russian influence, intimida-

[167] Discussions with Bulgarian and Romanian officials and analysts, Sofia and Bucharest, June 20–24, 2016.

[168] Discussion with Bulgarian analyst, Sofia, June 22, 2016.

tion, and subversion. For example, because of Moldova's and Serbia's weaker abilities to enforce the rule of law, Russia has significantly more freedom of action and capabilities there than in other countries in the region. Simultaneously, Russia likely has greater motivation to exercise control in these countries to prevent continued enlargement of the EU and NATO. Russia's ability to gain influence also varies among countries in the region that are NATO or EU members. Because of Bulgaria's Slavic culture, historical relationship to Russia, and dependency on Russian energy, Russia has more freedom of action there than in neighboring Latin-derived, energy-independent Romania.

Importantly, Russia does not have a completely free hand even in non-NATO and non-EU countries. Russia faces limits on its power even in Serbia, which in many ways is the best-case scenario for Russian hostile measures because of its Slavic roots and antipathy toward NATO stemming from the Balkan conflicts and Kosovo war. Serbians realize that they depend on the EU's economic aid and economic market; so, although they might be favorably disposed to Moscow's overtures in theory, there are limits to how far this courtship can go. That said, U.S., EU, and NATO interests are more limited in non-EU and non-NATO countries.

Hostile Measures Might Be More of a Local Threat Than an International One in Southeastern Europe

Russian hostile measures clearly pose a local threat to many countries in the region. Corruption in Bulgaria and Moldova likely hurt these economies and might force Moldovans and Bulgarians to pay higher energy prices than they would otherwise, effectively funding Russian influence out of their own pockets. The reliance on Russian news sources in Moldova, Serbia, and Bulgaria means that these populations receive slanted news. Russian connections to Greek, Bulgarian, and Serbian politicians could be eating away at public confidence in public institutions.

That said, Russia has been unable to turn economic, cultural, and informational leverage into substantive external political outcomes to date—pushing these countries to actively veto EU sanctions against Russia or block NATO deployments in the region. Although some of these countries have been unwilling participants in efforts by the EU and NATO to counter Russian aggression, they have typically followed the rest of the alliance when push came to shove. Perhaps the most dangerous, plausible scenario is that Russia could use hostile measures to stir up tension in the Balkans and cause NATO's peacekeeping efforts in Kosovo and in Bosnia and Herzegovina to fail—but this remains a theoretical capability so far rather than a proven one. In sum, Russian hostile measures might pose more of a local—or country-level—threat than an international one, at least for the moment.

Countering Hostile Measures Requires Continued Whole-of-Government Regional Engagement

Even though Russian hostile measures might pose a local threat rather than an international one to NATO and the EU, the United States should remain engaged in the region and consider options for addressing Russian hostile measures. First, it is important to keep Russian hostile measures in check within EU and NATO members. U.S. and European involvement—particularly in terms of economic aid and help with institution-building—cannot prevent Russian hostile measures in the region but has proven to be key in keeping the Russian influence in check. Analysts in Bulgaria and Romania were frustrated that U.S. assistance supporting civil society, media, and political parties ceased or greatly declined when these countries entered the EU. Given the increased challenges facing the EU because of Brexit (under which the United Kingdom will be first European Union member to leave the union) and radicalism, greater U.S. support to counter Russian influence through strengthening democratic institutions and building resilience might be valuable. U.S. military presence and exercises in the region might also have a deterrent value against even nonmilitary Russian actions.

Second, the United States has an interest in maintaining peace and security in the region. Southeastern Europe features several longstanding frozen conflicts and peacekeeping operations in the Balkans, Moldova, and elsewhere. In many of these cases, Russia demonstrated an interest in these situations and the ability to influence them. Although it is improbable that any one of these conflicts would reignite in coming years, ignoring them could strengthen Russia's hand and increase the risk of renewed violence. Crisis elsewhere in Europe might give Russia an incentive to provoke violence in Southeastern Europe, undermining decades of U.S. effort to create security and prosperity in the region. U.S. economic and political support, combined with military efforts in certain cases, will be critical for the U.S. policy of making Europe "whole, free, and at peace." U.S. military operations—often undertaken by the U.S. Army in such non-NATO countries as Bosnia, Moldova, Macedonia, or even Serbia—can improve local capacity and facilitate the long-term aspirations of these countries to eventually join the EU and NATO.

CHAPTER FOUR

Hostile Measures of Influence Elsewhere in Europe

Russian hostile measures in the Baltics and Southeastern Europe have attracted considerable attention over the past several years, but Russian hostile measures against the rest of Europe are also on the rise. Officials from European intelligence agencies, such as Germany's Office for the Protection of the Constitution and Sweden's Säpo intelligence agency, have emphasized the risks of Russian hostile measures, noting that Russian espionage and "influence operations" have increased dramatically of late.[1] In 2013, the Czech Republic's Security Information Service reported that the number of Russian intelligence operatives posing as diplomats, tourists, experts, academics, and entrepreneurs in the country was "extremely high." The service concluded, "The Cold War and the Soviet Union might have passed, but the same is not true for Russia's passion for trying to gain influence and taking active measures (such as the use of agents) to achieve this."[2]

Consequently, the issues become what exactly these Russian agents of influence can do and how big a threat they pose to European and U.S. interests over the next five years. To study these questions, this chapter looks at Russia's motivations, opportunities, and means to employ hostile measures. First, it looks at Russia's motiva-

---

[1] Elisabeth Braw, "Russian Spies Return to Europe in 'New Cold War,'" *Newsweek*, December 10, 2014.

[2] Matthew Day, "'Extremely High' Number of Russian Spies in Czech Republic" *The Telegraph*, October 27, 2014.

tions to use hostile measures in Europe and concludes that many Russian strategic objectives could—at least in theory—be served by employing hostile measures, with Western Europe being a more valuable target than Central Europe. Second, the chapter examines the structural factors shaping Europe in the short to medium term and argues that Europe's host of other problems—economic discontent, political disenfranchisement, massive immigration, and Islamic terrorism—will collectively present significant opportunities that Russia can exploit with hostile measures. Third, it examines the tools Russia has available to conduct hostile measures and argues that some, such as the growth of extremist parties, offer more potential than others, such as corruption, information operations, and shows of force. Ultimately, this chapter presents a mixed finding: Although macroconditions might make Europe more vulnerable to Russian hostile measures in coming years, Russia will also face significant constraints on its ability to use these tools effectively.

Motives

As previously discussed, Russia employs soft strategies and might not tie each investment to a specific objective. Still, it is worth briefly considering Russia's objectives in Europe at large and whether they could be served by using hostile measures. On the surface, many of Russia's stated objectives could be accomplished in part using hostile measures to undermine EU and NATO solidarity, to further its own security by weakening NATO (its longtime adversary) and more generally, to bolster its influence across region. A more nuanced look, however, suggests that Russia might have different motivations to employ hostile measures in the two regions: Western Europe might be a target for reasons in and of themselves while Russia might view Central Europe simply as a means to shape Western Europe, NATO, and Europe more broadly.

Influence in Western Europe still presents the largest strategic prize outside of the United States. For all the talk of "new Europe," Western Europe still comprises the lion's share of European economic wealth and military might.[3] In 2015, the United Kingdom spent $65.5 billion on defense, France spent $52.7 billion, and Germany spent $43.8 billion. Although still a far cry from U.S. military investments ($569.3 billion), these countries' outlays were more than quadruple that of the largest Central European state—Poland ($12.2 billion).[4] Western Europe also hosts several critical U.S. bases necessary for practically any U.S. operation in Europe—from the European Command Headquarters in Stuttgart to Landstuhl Regional Medical Center in Germany (the largest U.S. medical center outside the United States). Although the Army is prepositioning some equipment farther east in the Baltic states—along with Poland, Bulgaria, Romania, and Germany—U.S. supply lines in most European conflict scenarios with Russia will run from ports in the West overland via rail and road networks to the East.[5] Beyond the military calculations, Western Europe also has political importance, controlling two permanent UN Security Council seats (France and the United Kingdom) and accounting for much of the transatlantic trade.

As a result, if Russia denies the United States or NATO the political, military, and economic support of a handful of major Western European countries, it could inflict a serious blow to U.S. interests. Ultimately, this might lower Russia's bar for success in Western Europe: Russia does not necessarily need to install Russophile governments in the major Western capitals to achieve strategic success. Rather, anti-U.S. or even neutral, pacifist governments in Western Europe might be sufficient to help advance Russian interests by dividing some of the most powerful and influential countries within NATO and curbing global U.S. influence.

Many Central European countries are, by comparison, militarily less significant, economically poorer, and politically less influential. Outside of Western Europe, military spending drops off sharply, to the

---

[3] See Raphael S. Cohen and Gabriel M. Scheinmann, "Can Europe Fill the Void in U.S. Military Leadership?" *Orbis*, Vol. 58, No. 1, 2014, pp. 50–51.

[4] Ashely Kirk, "What Are the Biggest Defence Budgets in the World?" *The Telegraph*, October 27, 2015.

[5] Michelle Tan, "Army Wants to Double Tanks, Boost Soldiers in Europe," *Army Times*, July 15, 2015.

point of insignificance. Basing and transit rights might be strategically important depending on the scenario, but these countries will likely not be major force providers in any regional conflict in absolute terms. Although Western Europe is home to three of the largest economies in the world (Germany, the United Kingdom, and France) and five of the top 20 (Spain and Italy), the largest central European economy (Poland) ranks only 23rd, and many of the other countries do not even break the top 50.[6] Politically, these smaller countries are more important because the EU operates by consensus—as does NATO's highest authority, the North Atlantic Council—giving all members, regardless of size, the ability to wield veto power over decisions.[7] In practice, however, smaller states might not want to jeopardize their political and economic relationships with larger, more-powerful European states—let alone the United States—by blocking significant decisions.

In sum, Russia has plenty of motivation to pursue hostile measures over the near term. Many of Russia's stated objectives—undermining NATO, establishing itself as a great power, curbing U.S. power—could be accomplished through the application of hostile measures in Western and Central Europe. At least in theory, Western European countries are more-attractive targets than many Central European ones, in the sense that successfully influencing a major Western European state to take a more pro-Russian stance (or at least one that is anti-NATO, anti-EU) could possibly achieve Russian aims more directly and significantly. We now consider Russia's opportunity to use hostile measures against the Western and Central European countries.

Opportunities

As previous chapters illustrate, hostile measures are often most successful when they exploit preexisting opportunities in the system.

[6] See Central Intelligence Agency, "GDP (Purchasing Power Parity)," *World Factbook*, undated.

[7] NATO, "NATO's Assessment of a Crisis and Development of Response Strategies," June 16, 2011.

Consequently, this section examines the macroeconomic and political factors that will shape Western and Central Europe in the coming years to evaluate Europe's vulnerability to Russian hostile measures. Ultimately, Russia could be able to capitalize on four conditions—economic discontent, the unpopularity of the EU, the immigration crisis, and the rising terrorism problem.

Economic Conditions

First, economics shape Russia's ability to use hostile measures against the West in multiple ways. As will be discussed later, Europe's dependence on Russian goods and resources, perhaps most notably in the energy sector, can give Russia coercive power. Economic conditions can also play a less direct role in determining whether Russia can employ hostile measures successfully. Lackluster economic conditions can spur popular discontent and enable the rise of extremist parties that Russia can then leverage to its own ends. Weak economies can also lessen the resolve of countries to follow through with sanctions—given that sanctions can hurt the sender, as well as the target. Projecting into the next five years, overall economic conditions generally cut against Russia's ability employ hostile measures successfully in Western and Central Europe but not decisively so.

The European Commission projected "a mild recovery in the euro area surrounded by risks."[8] After significant setbacks during the euro crisis, the Commission projected that GDP inside the euro zone will continue to grow, from 1.6 percent in 2015 to between 1.7 percent and 1.9 percent in 2017.[9] That said, Europe's growth rate will still be considerably slower than other parts of the world. The Commission estimated global GDP growth at 3.2 percent in 2015, 3.6 percent in 2016, and 3.8 percent in 2017.[10] Europe's GDP growth was also lower

---

[8] European Commission Directorate-General for Economic and Financial Affairs, *European Economic Forecast, Winter 2016*, Institutional Paper 020, 2016. p. 1.

[9] European Commission Directorate-General for Economic and Financial Affairs, 2016, p. 2.

[10] European Commission Directorate-General for Economic and Financial Affairs, 2016, p. 2.

than other parts of the developed world. According to World Bank estimates, the U.S. GDP grew at 2.5 percent in 2015 and will grow by 2.7 percent in 2016 and 2.4 percent in 2017.[11]

Europe's labor market also presents a similarly mixed finding. On the one hand, the European Commission projects the euro area's unemployment rate will decline from 11 percent in 2015 to 10.5 percent in 2016 and 10.2 percent in 2017.[12] Unemployment rates vary by country, but most countries' job situations will improve.[13] That said, whether these gains will turn into increased economic satisfaction—and, by extension, more resiliency to Russian hostile measures—remains debatable. First, despite the generally improving conditions, unemployment will likely remain in the double digits through 2017 in several countries: Spain (18.9 percent), Croatia (13.8 percent), Cyprus (13.2 percent), Italy (11.3 percent), Portugal (10.8 percent), and France (10.3 percent).[14] Second, although the labor market in most of the euro zone will improve, the unemployment rate is expected to tick upward in Germany (from 4.8 percent in 2015 to 5.2 percent in 2017) and Austria (from 6 percent in 2015 to 6.4 percent in 2017).[15] Taken together, many of the countries that currently have surging left- or right-wing political parties will likely continue to face either high or rising unemployment for the near future.

On balance then, the economy in Europe will likely improve and make it somewhat more resilient to any potential Russian attempts to leverage economic conditions to its advantage. Three caveats are in order, however. First, European economies will grow only modestly, so popular discontent—to the extent they stem from economic conditions—might

---

[11] World Bank, "United States," undated-b.

[12] European Commission Directorate-General for Economic and Financial Affairs, 2016, p. 6.

[13] European Commission Directorate-General for Economic and Financial Affairs, 2016, p. 6.

[14] European Commission Directorate-General for Economic and Financial Affairs, 2016, p. 1.

[15] European Commission Directorate-General for Economic and Financial Affairs, 2016, p. 1.

not go away any time soon. Second, some parts of Europe will benefit more than others—with Southern Europe still experiencing relatively high unemployment for the foreseeable future. Finally, as the Commission's report notes, these economic forecasts remain highly uncertain and subject to global economic trends. If the global economy were to suffer, Europe's modest gains might also decline as a result.

Even if Europe's economy improves, Russia can still capitalize on other political crises and on dissatisfaction with the EU. According to Fall 2015 European Commission polling, 54 percent of Europeans surveyed said their voices do not count in the EU.[16] Indeed, at no point from 2004 on has a majority of Europeans surveyed in the biannual polls reported that their voices count.[17] Overall, the percentage of Europeans with a positive opinion of the EU declined from about 50 percent in 2004 to 37 percent in 2015, although this is somewhat improved from the low of 30 percent in late 2012 and early 2013.[18] More troubling, perhaps, is that dissatisfaction tends to be concentrated in certain states: A plurality of respondents of the United Kingdom, Austria, the Czech Republic, and Cyprus all had negative views of the EU.[19] Pluralities of German, Czech, British, French, Austrian, Greek, and Cypriot respondents were pessimistic about the EU's future.[20] This dissatisfaction increases the ability of outside actors, such as Russia, to try to shatter European unity. Already, the United Kingdom has voted itself out of the EU, and other countries might choose to follow suit.

Immigration Crisis

Intertwined with dissatisfaction with the EU is Europe's ongoing immigration crisis. Indeed, according to the European Commission poll, immigration was the top issue in every EU country except Portugal,

---

[16] European Commission, Directorate-General for Communication 2016, p. 9.

[17] European Commission, Directorate-General for Communication 2016, p. 9.

[18] European Commission, Directorate-General for Communication 2016, p. 6.

[19] European Commission, Directorate-General for Communication 2016, p. 7.

[20] European Commission, Directorate-General for Communication 2016, p. 12.

where it came in a close second.[21] There is good reason for this concern. Thanks to the ongoing civil war in Syria and the turmoil throughout the Middle East and North Africa, Europe has been flooded with migrants. Some 1,321,560 people, mostly from Syria, applied for asylum in the EU (or roughly 260 immigrants per 100,000 residents) and this number likely undercounts the number of migrants because not all of them applied.[22] Reportedly, 476,000 asylum seekers went to Germany (or roughly 587 immigrants per 100,000 residents), but German officials estimate that the true number of migrants in the country could be as many as a million.[23] Other countries have comparatively fewer migrants in absolute terms but considerably higher relative to their population: Hungary has 1,799 asylum seekers to every 100,000 people, Sweden has 1,667, and Austria has 1,027.[24]

Islamic Terrorism

Related to the immigration crisis, Western and Central Europe also face the threat of Islamic terrorism. According to some estimates, as many as 6,000 Europeans have gone to fight the Islamic State in Syria and Iraq, including some 1,700 from France and about 760 apiece from Germany and the United Kingdom.[25] These fighters could turn their fighting skills and experience against their home countries when they return from the battlefield. Already, Islamic terrorism is on the rise in Europe. According to the *New York Times*, there were 20 Islamic State–linked attempted and successful terrorist attacks in Europe from December 2014 to March 2016—including in Belgium, Bosnia and Herzegovina, Denmark, France, Germany, and Turkey.[26] With no end

---

[21] European Commission, Directorate-General for Communication 2016, p. 15.

[22] "Migrant Crisis: Migration to Europe Explained in Seven Charts," BBC News, March 4, 2016.

[23] "Migrant Crisis: Migration to Europe Explained in Seven Charts," 2016.

[24] "Migrant Crisis: Migration to Europe Explained in Seven Charts," 2016.

[25] Ashley Kirk, "Iraq and Syria: How Many Foreign Fighters are Fighting for ISIL?" *The Telegraph*, March 24, 2016.

[26] Karen Yourish, Tim Wallace, Derek Watkins, and Tom Giratikanon, "Brussels Is Latest Target in Islamic State's Assault on West," *New York Times*, March 25, 2016.

in sight for the Middle East turmoil or the Islamic State, the threat of Islamic terrorism will likely continue and could increase over the next five years.

Ultimately, Europe's lackluster economic recovery, EU disenfranchisement, immigration, and terrorism problems make Europe ripe for Russia's efforts in several ways. First, on the most basic diplomatic level, these issues allow Russia to argue that Europe should focus on its own economic recovery and "the southern threat" (the destabilization of the Middle East and North Africa) rather than on such issues as Ukraine. Moreover, Russian officials can argue that Russia can play a constructive role in the Middle East conflicts that fuel Europe's destabilization. Indeed, after the Islamic State–inspired terrorist attacks in Paris, French President Francois Hollande met with Putin to discuss counterterrorism and Middle East policy, raising the possibility of a warming in relations between the two countries.[27]

Second, the economic, immigration, and terrorism issues offer Russia an opportunity to divide NATO and further its own great power status. Europe's economic problems pit Germany and Northern European countries against the more fiscally unstable Southern European countries. Similarly, by the nature of geography, much of Southern Europe—Italy, the Balkans—is affected more immediately and far more extensively by the immigration crisis than by Russia's adventures in Ukraine. Arguably, the same is true for Western European states—e.g., France and Germany—that have substantial Islamic minorities but are buffered from Russia by Central European states. In contrast, other states—such as the Baltics and Poland—might be more insulated from the immigration issues but are more immediately threatened by Russia. Given the relatively low support for the EU to unite the continent, this difference in strategic priorities can create rifts that Russia can exploit to its own ends.

Third, as will be discussed, economics, immigration, and terrorism fuel the growth of extremist parties that run an economic nationalist, anti-immigration, anti-EU and pro-

---

[27] Andrew Osborn, "Paris Attacks, Hollande Visit May Spur Kremlin Push to End Isolation," Reuters, November 18, 2015.

hardline security platform. And as will be explained later in this chapter, many of these parties also take a more lenient position toward Russia. In other words, if the EU remains unpopular and immigration and terrorism remain concerns, there might be an increased chance of parties that are more favorably disposed to Russia winning elections in Western and Central European states.

Means

Given that Russia certainly will have the motivation and the opportunity to employ hostile measures against the West in the coming years, this raises the question: Which tools will Russia have at its disposal? Perhaps, a good place to start is what tools it currently uses to shape Western Europe. Of course, this does not constitute an exhaustive list—some might not be publicly known and Russia could always develop new forms of leverage over the next half-decade. Still, evaluating what Russia uses currently can help define the scope of plausible Russian actions and a method for critically evaluating the impact of each tool.

Political Influence

Rather than buying favor with individual leaders, Russia also might try to gain influence in the West by funding political parties themselves. Casting itself as the defender of conservative values, Russia already has inroads into the far right of the European spectrum. In March 2015, 150 representatives of far-right political parties—including the British National Party and the German neo-Nazi NPD—met in St. Petersburg to coordinate policy and to bash Western support for the Ukrainian government.[28] And yet, according to media reports, Russian ties to the far right extend beyond simply hosting conferences to actually bankrolling their operations.[29] As listed in Table 4.1, Russia is said to have ties—of differing types and strengths—to right-wing populist parties across Europe.

[28] "Europe Far-Right Parties Meet in St Petersburg, Russia," BBC News, March 22, 2015.

[29] See Foster and Holehouse, 2016.

Russia's most overt tie is perhaps to Marine Le Pen's Front National in France. In November 2014, Le Pen, the daughter and successor to her father, Jean-Marie Le Pen, as leader of this Euro-skeptic French party, received the first €9 million of an alleged €40-million loan from the First Russian-Czech bank, supposedly to help bankroll her presidential bid in 2017.[30] Le Pen previously declared her admiration for Putin, support for the Russian stance on Ukraine, and condemnation of the French decision not to supply Russia with its two *Mistral* warships built in French dockyards.[31] The hacker group Anonymous International later claimed to intercept texts from Kremlin insiders suggesting that Le Pen's stance recognizing Crimea was in conjunction with

Table 4.1
Russia's Ties to Right-Wing Populist Groups

Country	Names of the Political Party
Austria	Austria Freedom Party (FPÖ)
Belgium	*Vlaams Belang*
France	Front National
Germany	National Democratic Party AfD
Great Britain	British National Party United Kingdom Independence Party (UKIP) (suspected)
Hungary	*Jobbik* *Fidesz*
Italy	Northern League Forza Italia
Slovakia	People's Party–Our Slovakia
Poland	*Nowa Prawica*

SOURCE: "In the Kremlin's Pocket: Who Backs Putin, and Why," *The Economist*, February 14, 2015.

---

30 David Chazan, "Russia 'Bought' Marine Le Pen's Support over Crimea," *The Telegraph*, April 4, 2015.

31 Gianluca Mezzofiore, "Marine Le Pen's Front National Borrows €9m from Russian Lender," *International Business Times*, November 24, 2015.

Russian authorities—possibly at their behest.[32] For her part, Le Pen denies Russian money influenced her positions and claims that she turned to the Russian bank after being denied loans in France.[33]

Marine Le Pen is not alone, however. In Germany, Russia is said to have ties both to the neo-Nazi New Democratic Party and the euroskeptic AfD.[34] Of the two, Russia's relationship with the *Alternative für Deutschland* (AfD)—which carried 13 percent of the vote to become Germany's third-largest party in the 2017 parliamentary elections—is more significant because it threatens Chancellor Angela Merkel's ruling center-right Christian Democrats' hold on power.[35] In a "Thesis on Foreign Policy" posted on AfD's website, one of its leading politicians, Alexander Gauland, called for a more cooperative relationship with Russia and likened Russia's relationship with Ukraine and Belarus to Germany's relationship with the cities of Cologne and Aachen.[36] In November 2014, AfD representatives met with Russian Ambassador to Germany Vladimir Grinin, who allegedly provided "strategic advice."[37] According to media accounts, the party also has cohosted seminars on "Migration as a Destabilizing Element" with the Russian embassy.[38]

---

[32] Lucy Draper, "Hackers Leak Messages 'Between Kremlin and France's Front National,'" *Newsweek*, April 3, 2015; Chazan, 2015.

[33] Chazan, 2015.

[34] Melanie Amann, Markus Becker, Benjamin Bidder, Hubert Gude, Konstantin von Hammerstein, Alexej Hock, Christiane Hoffmann, Martin Knobbe, Peter Maxwill, Peter Müller, Gordon Repinski, Sven Röbel, Anna Sadovnikova, Matthias Schepp, Jörg Schindler, and Christoph Schult, "The Hybrid War: Russia's Propaganda Campaign Against Germany," Spiegel Online International, February 5, 2016.

[35] Damien McGuinness, "Germany Jolted by AfD Right-Wing Poll Success," BBC News, March 14, 2016b; Kate Connolly, "German Election: Merkel Wins Fourth Term but Far-Right AfD Surges to Third," *The Guardian*, September 24, 2017.

[36] Alexander Gauland, "Thesenpapier Außenpolitik [Thesis on Foreign Policy]," *Alternative für Deutschland*, September 10, 2013.

[37] Elisabeth Braw, "Putin Seeks to Influence Radical Parties in Bid to Destabilise Europe," *Newsweek*, January 9, 2015a. For its part, AfD admits to the meeting but denies the claim that it receives instructions from Russia.

[38] Amann et al., 2016.

Whether the AfD's relationship extends beyond shared sympathies, however, is debatable. The AfD partially financed itself by selling small gold bars and old deutschmark coins, both to raise a profit and to maximize the matching state subsidies to the party.[39] The German newspaper *Bild* reported that a Moscow-based Russian think tank, the Centre for Strategic Communications, suggested that Russia could fund the AfD by helping supply it with gold.[40] For its part, AfD denies getting money or guidance from Russia.[41] The allegations were never proven and Germany later closed the funding loopholes that made such a venture profitable.[42]

Comparable stories play out with other far-right Western Europe parties. In Austria, Heinz-Christian Strache, the leader of FPÖ, visited Moscow in 2015 and advocated the end of sanctions against Russia.[43] He also called on the EU to "stop playing the stooge of the U.S. in the encirclement of Russia."[44] The FPÖ has also been accused of taking Russian money, a claim Strache denies.[45] Other Western European right-wing parties with suspected ties to Russia include Poland's Nowa Prawica and Britain's UKIP and British National Party. Belgium's Vlaams Belang and others have been accused of receiving funds from Russia—although these allegations have not been proven.[46]

---

39 Assuming a party earned more than 0.5 percent in the last election, Germany provided a subsidy to the party of $0.92 per vote up to the value of the income from parties' other revenues. As a result, even if the gold fundraiser was not particularly profitable in and of itself, it helped maximize the state subsidy of the AfD. Germany closed the loophole in 2015. Ben Knight, "After the Gold Rush: AfD Loses State Subsidies," *Deutsche Welle*, December 18, 2015.

40 Tony Patterson, "Putin's Far-Right Ambition: Think-Tank Reveals How Russian President Is Wooing—and Funding—Populist Parties Across Europe to Gain Influence in the EU," *The Independent*, November 25, 2015.

41 Braw, 2014.

42 Knight, 2015.

43 Braw, 2015a; Peter Baker and Steven Erlanger, "Russia Uses Money and Ideology to Fight Western Sanctions," *New York Times*, June 7, 2015.

44 Mezzofiore, 2015.

45 Mezzofiore, 2015; Baker and Erlanger, 2015.

46 Braw, 2015a.

Further east, Russia developed friendly ties with both of Hungary's right-wing parties—the ruling Fidesz and the Jobbik. Orbán, the current prime minister and leader of the Fidesz party, started his career after the fall of Communism as a pro-Western liberal, but as he has grown more authoritarian, he has become increasingly anti-EU and sympathetic to Russia.[47] In November 2011, Orbán said that an "Eastern Wind" was blowing in the world, although he added that "we're sailing under a Western flag."[48] In July 2014, Orbán argued that Russia—along with other authoritarian countries—should serve as the model for Hungary.[49] More recently, on the occasion of Putin's 2015 visit, he said: "We are convinced that the isolation of Russia from Europe is not feasible."[50]

Fidesz's principal competition is the Jobbik party, which leans farther right and also has Russian ties. Russia is rumored to have provided Jobbik with funds and organizational assistance.[51] More concretely, in October 2015, the European Parliament's Legal Affairs Committee suspended the parliamentary immunity of Jobbik MEP Béla Kovács after discovering he regularly met Russian diplomats covertly and paid monthly visits to Moscow.[52] A Hungarian newspaper reported that Kovács's wife, Russian Svetlana Izstosina, was formerly employed by the KGB as a messenger and Kovács was suspected of serving as an accomplice.[53]

---

47 Interview with think tank analyst, Cambridge, United Kingdom, February 3, 2016.

48 "Orbán: Keleti szél fúj [Orbán: Eastern Wind Is Blowing]," *Index* [Budapest], November 5, 2011.

49 "Orbán: Keleti szél fúj [Orbán: Eastern Wind Is Blowing]," 2011.

50 Rick Lyman and Helene Bienvenufeb, "Hungary Keeps Visit by Putin Low-Key as It Seeks to Repair Relations with West," *New York Times*, February 17, 2015.

51 Foster and Holehouse, 2016; interview with think tank analyst in Cambridge, United Kingdom, February 3, 2016.

52 "Jobbik MEP Béla Kovács, Accused of Spying for Russia Previously, to Lose Immunity," *Hungary Today*, October 13, 2015.

53 Damien Sharkov, "Far-Right MEP Accused of Acting as Russian Spy," *Newsweek*, September 26, 2014.

Similar patterns play out elsewhere in Central Europe. In recent Slovak national elections, the People's Party Our Slovakia won 8 percent for the first time. The party's leader, Marian Kotleba, occasionally wears uniforms reminiscent of the Nazi-occupied Slovak government and espouses a positive view of Russia, opposing Western-backed sanctions.[54]

Although most of the popular attention centers on Russia's influence on right-wing populist groups, there is also potential affinity between Russia and far-left parties, including Podemos in Spain and Die Linke in Germany.[55] Founded in 2014, Podemos' opposition to austerity measures helped the party capture 21 percent in national elections in 2015, making it the third-largest party in parliament.[56] Podemos' secretary general, Pablo Manuel Iglesias Turrión, a 37-year-old former political science professor, has close ties with Russian-friendly Greek Prime Minister Alexis Tsipras.[57] Although Iglesias does not have any reported formal ties with Russia, he shares Tsipras' sympathies for Putin's position. He criticized Europe's decision to sanction Russia, as well as Europe's support for the Maidan revolution, by noting that "it was unreasonable to back what—to use a softer expression than *coup d'etats*—was an illegal displacement of political power."[58] When the Spanish newspaper *El País* asked Iglesias in September 2015 about his pro-Russian views and noted that leaders of such left-wing opposition parties as Podemos would be jailed in Russia, Iglesias responded, "I agree. In Russia we would end up in prison, but perhaps in the United States we would be delivered a few blows as well."[59] Importantly, there

---

[54] Leonid Ragozin, "Putin's Hand Grows Stronger as Right-Wing Parties Advance in Europe," Bloomberg News, March 15, 2016.

[55] "Stirring the Pot," 2015.

[56] Ashifa Kassam, "Spanish Election: National Newcomers End Era of Two-Party Dominance," *The Guardian*, December 21, 2015.

[57] Giles Tremlett, "The Podemos Revolution: How a Small Group of Radical Academics Changed European Politics," *The Guardian*, March 31, 2015.

[58] Tremlett, 2015.

[59] John Carlin, "Pablo Iglesias: "Catalans, Stay with Us and Let's All Kick Out Rajoy Together," *El País*, September 24, 2015.

are no reported allegations of direct Russian funding of Podemos thus far, although there have been accusations of Russia's allies, Venezuela and Iran, funding Podemos' leaders—a claim Podemos vehemently denies.[60]

Russia's link with the German Die Linke party (the successor of East Germany's Communist Party) is more clearly documented. Although the party fell out of favor after the fall of Communism, it is experiencing a political resurgence. In 2014, for the first time in 25 years, a Die Linke politician became minister-president of one of Germany's states, the central state of Thuringia.[61] Like other far-left parties, Die Linke advocates a more conciliatory tone toward Russia. The party proposed dissolving NATO and replacing it with a "collective security system" with Russia as a member.[62] Die Linke leader Gregor Gysi condemned Russia's annexation of Crimea—but he also pushed for compromise with Russia and, like the far-right parties, sent election observers to Crimea and Donetsk.[63]

Arguably, populist European parties pose one of the most serious challenges because these parties command significant vote shares across the continent. In the March 2016 state elections in Germany, AfD earned double-digit shares of the electorate threatening Chancellor Angela Merkel's ruling center-right Christian Democrats.[64] In 2014, Marine Le Pen's Front National claimed about 25 percent of the votes in France and although the party secured only eight of 577 seats in the National Assembly in 2017, Le Pen still earned 34 percent of

---

[60] See Stephen Burgen, "Podemos Leaders Deny Venezuela Government Funding Link Claims," *The Guardian*, April 6, 2016; Martin Deflin, "Podemos and the Iran-Venezuela Connection," Deutsche Welle, January 27, 2016.

[61] Derek Scally, "Election of Die Linke State Premier Causes Stir in Germany," *Irish Times*, December 5, 2014.

[62] "What Does the Left Party Want for Europe?" *The Local*, May 14, 2014; Die Linke, "Programme Of the DIE LINKE Party," December 2011, p. 70.

[63] Gysi, Gregor, "Ukraine—Diplomacy Is the Only Way," *The Bullet*, Socialist Project E-Bulletin, No. 951, March 19, 2014; Baker and Erlanger, 2015.

[64] McGuinness, 2016b.

the vote in the run-off presidential election that year.[65] In the United Kingdom, UKIP won 12.6 percent of the vote in 2015 (although this was a smaller share than the 27 percent that it earned previously).[66] In 2015, Austria's FPÖ commanded 30.4 percent of the vote.[67] And in Hungary, Jobbik scored victories in by-elections in 2015.[68]

In all likelihood, Russian influence is not the most important reason for these parties' success. As already mentioned, these parties draw on Europe's ongoing economic problems, the refugee crisis from North Africa and the Syrian civil war, dissatisfaction with the EU, and a host of other factors. Still, whatever the reason for their rise, these parties' increased political power and ties to Russia will present significant opportunities for Russian strategy in years to come.

To an extent, Russia already uses right-wing parties for political legitimacy. When Crimea held a referendum in 2014 on whether it should join Russia, the pro-Russian, Belgium-based Eurasian Observatory for Elections and Democracy organized an election-monitoring team, which included representatives from the FPÖ, Jobbik, and Vlaams Belang. Unsurprisingly, the team pronounced the elections free and fair.[69] The Observatory had already performed similar election-monitoring functions with other Russian-backed separatist states—Abkhazia, Nagorna-Karabakh, and Transnistria—and it likely will continue to do so in the future.[70] As these right-wing parties become more politically significant, their "stamp of approval" of Russian actions might carry even more weight than they do today.

---

[65] Kim Willsher, "France's Far-Right National Front Basks in Election Victory," *Los Angeles Times*, May 27, 2014; Fabio Benedetti Valentini, "Le Pen Seeks to Revive France's National Front with Name Change," Bloomberg Politics, January 7, 2018.

[66] "Election 2015: Results," BBC News, undated; Willsher, 2014.

[67] Shadia Nasralla, "Austrian Far-Right Party Gets Electoral Boost from Migrant Crisis," Reuters, September 27, 2015.

[68] Zoltan Simon, "Hungary Radical Party Wins By-Election in Breakthrough Vote," Bloomberg, April 12, 2015.

[69] Andrew Higgins, "Far-Right Fever for a Europe Tied to Russia," *New York Times*, May 20, 2014a.

[70] Orenstein, 2014.

Right-wing populist parties could also pose less-direct problems. For example, right-wing parties often call for stricter controls on immigration. Currently, many European governments—including Austria, Hungary, Slovenia, and Slovakia—are walling off their borders.[71] If as one European analyst from that region conjectured, these countries are successful in this effort and the flow of migrants does not subside, it could lead to refugees "getting stuck" in formerly volatile areas, such as the Balkans, and reigniting old conflicts.[72] If this were to happen, the United States would face pressure to shift resources away from Russia to handle these challenges.

Ultimately, what makes the right-wing parties such a potentially attractive tool for Russian hostile measures and so problematic for U.S. interests over the next five years is that they require little in the way of Russian encouragement. The right-wing populist groups likely would still oppose the EU, NATO, and the United States even without Russian influence, and they might still be electorally successful even without Russian help. Additional Russian aid, however, could exacerbate an already existing problem.

Economic Leverage

As previously discussed, the macroeconomic conditions in the coming years generally favor the West over Russia—Europe's economy is projected to gradually recover while Russia's will likely stagnate. Nonetheless, Russia still has several economic levers it can employ against Western and Central Europe, most notably with Europe's dependence on Russian energy. However, Russia's ability to use Europe's energy dependence is both variable across countries and in overall decline as Europe pursues diversification and Russia continues to rely heavily on energy exports for revenue.

Europe imports much of its energy. According to the European Commission, the EU imports about 53 percent its energy—including 90 percent of its crude oil, 66 percent of its natural gas, 42 percent of

---

71 Tom Batchelor, "The New Iron Curtains: Where the Fences Are Going Up Across Europe to Keep Migrants Out," *Express*, December 6, 2015.

72 Interview with think tank analyst in Cambridge, United Kingdom, February 3, 2016.

its coal and other solid fuels, and 40 percent of its uranium and other nuclear fuels—all at a cost of more than €1 billion per day.[73] Much of Europe's oil—and even more importantly, its natural gas—comes from Russia. In 2014, Russia exported approximately 140 billion cubic meters of natural gas to Europe—or about 30 percent of the EU's total imports, although this was down from 39 percent the previous year.[74] Moreover, Russian energy companies have ties—and sometimes, controlling stakes—in Central and Western European energy companies.[75] The concern is that Russia could threaten to cut Europe off if these countries do not back its policies.

This fear becomes more acute looking at individual European countries. In 2012, the Baltics and Finland were almost entirely dependent on Russian natural gas exports—although, as already discussed, the Baltics' dependence is declining and gas is a relatively small percentage of overall energy use. Central Europe also relied heavily on Russian gas exports: Hungary and Slovakia got more than 80 percent of their natural gas from Russia; the Czech Republic, Turkey, Austria, Poland, and Slovenia also all imported more than half of their natural gas from Russia.[76] Indeed, some officials in these countries claim—rightly or wrongly—that they are forced to take a softer line on Russia partly because their populations rely on Russian energy to heat their homes and cook their food.[77]

Central Europe already tried to increase its resilience to potential Russian manipulations of the energy market. Beginning in 2014, it conducted a series of "stress tests," preparing for the possibility that Russia would either stop its European gas exports or disrupt the flow

---

[73] European Commission, "Imports and Secure Supplies: Diverse, Affordable, and Reliable Energy from Abroad," March 31, 2016b.

[74] Qishloq Ovozi, "The European Union, The Southern Corridor, and Turkmen Gas," Radio Free Europe/Radio Liberty, April 23, 2015.

[75] For example, Russia's Gazprom and Austria's OMV have an almost 50-year old relationship. OMV, "OMV and Gazprom Celebrate 40 Years of Natural Gas Import from Russia into Austria," April 17, 2008.

[76] "Conscious Uncoupling," *The Economist*, April 5, 2014.

[77] Interviews with senior Hungarian government officials, Budapest, May 2014.

of gas through Ukraine.[78] Europe also attempted to diversify its energy imports. Poland opened its first seaborne LNG terminal in October 2015 and plans to ship gas elsewhere in Central Europe, including the Czech Republic and Slovakia.[79] Lithuania also opened its own LNG terminal, as already mentioned,[80] and there are plans for Greek LNG terminals to begin exporting to Bulgaria in Southeastern Europe.[81]

Despite these measures, Europe will remain partially dependent on Russian gas—at least in the short term. Although Europe can import more from Norway and other producers, this will likely be a more expensive option.[82] Moreover, Europe will need to contend with other buyers (such as China) for natural gas, and turmoil in the Middle East could threaten energy supplies. Indeed, even as Europe has urged energy independence from Russia, a Gazprom-led consortium with European oil companies pushed for building Nord Stream 2, designed to increase Russian natural gas exports to Western Europe (while bypassing Central Europe)—despite U.S. opposition.[83]

Even if Europe does remain dependent on Russian gas for the short term, it is not clear that Russia will be able to leverage such dependence effectively as a coercive tool. As much as Europe needs Russian energy, Russia needs European consumers to survive economically. According to Sergei Aleksashenko, a former deputy chairman of the Russian central bank in the late 1990s who is now a Brookings Institution economist, upward of 80 percent of Russia's economy comes from exporting raw materials and commodities, and Russia simply cannot find a substitute

---

[78] European Commission, "Energy Security Strategy," March 31, 2016a.

[79] Marek Strzelecki, "Poland Opens LNG Terminal, Pledges to End Russian Dependence," Bloomberg, October 12, 2015.

[80] Georgi Kantchev, "With U.S. Gas, Europe Seeks Escape from Russia's Energy Grip," *Wall Street Journal*, February 25, 2016.

[81] Kantchev, 2016.

[82] "Conscious Uncoupling," 2014.

[83] Radu Sorin Marinas, "U.S. Raises Fresh Concerns Over Gazprom-Led Nord Stream 2," Reuters, February 18, 2016.

for European markets.[84] Although Russia tried exporting more natural gas to China, the European and Chinese markets traditionally have been serviced by different oil fields—and as China's economy slowed, its demand for raw materials slackened. Moreover, Russia's other industries, such as arms (5 percent of GDP), agriculture (3 percent), and cars (2 percent), simply do not encompass enough of Russian economic output to compensate for a decline in natural resource exports.[85]

Finally, the poor state of Russia's economic health—at least in the short run—decreases the likelihood that Russia will opt for economic hostile measures in the short term. After a series of down years, Russia experienced modest growth in 2017 and the World Bank projects that Russian GDP will grow 1.4 percent through 2019.[86] If Russia were to use energy as a coercive tool however, this might upset these projections. Moreover, after declining oil prices, Russia's sovereign wealth funds have been depleted.[87] As these funds decrease, Russia's cushion to withstand the loss of revenues from energy sales to Europe will also erode.

In sum, Russia's ability to employ economic hostile measures against Central and Western Europe might be declining both because Europe now is moderately more energy independent and because Russia might need to sell to the European market as much as Europe needs to buy Russian energy. Although Russia still might try to use economic coercion on a smaller scale against individual countries, large-scale economic coercion against Europe will become progressively less likely and less effective.

---

[84] Dragan Stavljanin, and Ron Synovitz, "'Damn Lies, Deep Crisis' in Russian Economy, Says Former Central Banker," Radio Free Europe/Radio Liberty, February 16, 2016.

[85] Stavljanin and Synovitz, 2016.

[86] Andrey Ostroukh, "Russian Economy Seen Growing From 2017 Onwards: World Bank," Reuters, May 23, 2017.

[87] Evgenia Pismennaya and Anna Andrianova, "Russia's Economy Is Tanking, So Why Is Putin Smiling?" Bloomberg Business, February 29, 2016.

Personal Corruption

Exploiting personal corruption potentially offers Russia the most direct tool for accomplishing its objectives in Western Europe. Russia could buy senior political leaders and then use this influence to its advantage—perhaps not to change these Western governments' policies wholesale, but to nudge these leaders' countries in a pro-Russian direction, or at the very least throw up obstacles to U.S. or NATO action in Europe. Although there is anecdotal evidence of Russia—or more accurately Russian businesses—developing personal financial ties to a series of Western European leaders and of Western European leaders taking a softer line on Russia, there is no clear evidence on the unclassified level of a quid-pro-quo relationship, where agents of the Russian government paid Western leaders in exchange for taking a given policy position. Thus, it seems unlikely that personal corruption will prove decisive for Russia soon—at least in Western Europe.

Perhaps one of the most publicized alleged examples of Russia buying influence in Western Europe is that of Gerhard Schröder, the former leader of the Social Democratic Party in Germany and chancellor of Germany from 1998 to 2005. A critic of the United States and the Iraq War while in office, Schröder developed a close relationship with Putin, "putting relations between Berlin and Moscow on the friendliest terms since Nazi Germany fought the Soviet Union on the battlefields of World War II."[88] Shortly after leaving the chancellorship, Schröder took a job as the chairman of the joint German-Russian Nord Stream pipeline, majority-controlled by Gazprom.[89] According to media accounts, Schröder draws a salary of €250,000 for his efforts.[90] Even in 2005, a decade before the current Russian-Western tensions, the deal raised eyebrows. When media asked about the position, Gazprom responded that

[88] Craig Whitlock and Peter Finn, "Schroeder Accepts Russian Pipeline Job," *Washington Post*, December 10, 2005.

[89] Whitlock and Finn, 2005.

[90] Erik Kirschbaum, "Putin's Apologist? Germany's Schroeder Says They're Just Friends," Reuters, March 27, 2014.

"this position is not related to any kind of favor on our part" and not part of some underhanded deal.[91]

The story grows murkier, however, given the background of the pipeline's chief executive, Matthias Warnig. Warnig was a foreign intelligence officer in the East German Stasi.[92] He supposedly met Putin back in the early 1980s, when Putin was still a KGB officer.[93] After the Berlin Wall fell, Warnig renewed Putin's acquaintance in 1991, when the former was representing Dresdner bank and the latter was a civil servant in charge of trade. With Putin's help, Warnig opened Dresdner's branch in St. Petersburg and the two men became close personal friends after Warnig helped Putin's wife receive medical attention in Germany. Over the next quarter-century, Warnig's influence rose with Putin's star. Warnig later served on boards of directors of Russia's second-largest credit institution, VTB Bank (nicknamed "the bank of Putin's friends"), the Rosneft (Russia's largest oil company), and Transneft (the Russian company that owns the pipelines).[94]

Given Warnig's ties to Putin and his job offer to Schröder, there is a possible connection between Russia's employment of Schröder and Schröder's public position of likening Russia's intervention in Ukraine to the NATO intervention in Serbia and his pushing for a softer response by the West to Ukraine.[95] Despite these reports, however, there is no solid evidence in the public sphere tying Russian money directly to Schröder's political views. Other former German chancellors Helmut Schmidt and Helmut Kohl have also advocated a less confrontational stance toward Russia. And so, although Schröder clearly benefits from his relationship with Russia and advocates a softer line, there is no firm evidence that this was part of a deliberate attempt by the Russian government to wield influence by targeted corruption.

---

[91] Whitlock and Finn, 2005.

[92] Dirk Banse, Florian Flade, Uwe Müller, Eduard Steiner, and Daniel Wetzel, "Circles of Power: Putin's Secret Friendship with Ex-Stasi Officer," *The Guardian*, August 13, 2014.

[93] Whitlock and Finn, 2005.

[94] Banse et al., 2014.

[95] Kirschbaum, 2014.

Schröder is not the only prominent Western politician to be accused in the media of being on the Russian take. Disgraced former Italian Prime Minister Silvio Berlusconi, for example, also has a well-reported long friendship with Putin. In July 2015, Berlusconi stated that Putin offered him Russian citizenship and a post as economic minister,[96] and Berlusconi certainly has been outspoken in his defense of Putin. He accompanied Putin on a tour of Crimea, earning Ukraine's ire for drinking Crimea's oldest bottle of wine at a winery on the EU's sanction list.[97] Returning to Italy, he stated that Crimea's elections were democratic and recounted, "You should see the love, the gratitude, and the friendliness that welcomed Putin" and how "Women threw themselves into his arms saying, 'Thank you, Vladimir. Thank you, Vladimir.'"[98] As with Schröder, however, there is no hard evidence that Berlusconi's political views are direct result of Russian payments.

Further east, current Czech President Miloš Zeman's relationship with Russia also raised eyebrows. During the Kosovo intervention, when Zeman was prime minister, he referred to NATO and its supporters as "warmongers" and "primitive troglodytes who assume everything can be achieved by bombing."[99] After his 2013 election, he promised a "pragmatic" approach to Russia, enhancing "economic cooperation mechanisms."[100] Zeman referred to former Ukrainian Prime Minister Arseni Yatsenyuk as a "prime minister of war" and insisted that the "Maidan [movement] was no democratic revolution," but a "civil war" fought between rival "gangs."[101] As with Schröder

---

[96] Alice Phillipson, "Berlusconi Says Vladimir Putin Wants Him to Become Russia's Economy Minister," *The Telegraph*, July 23, 2015.

[97] Claire Bigg, "Ukraine Livid as Putin, Berlusconi Swig Crimea's Oldest Bottle of Wine," Radio Free Europe/Radio Liberty, September 17, 2015.

[98] Steve Scherer, "Italy's Berlusconi Says Crimea Split from Ukraine Was Democratic," Reuters, September 27, 2015.

[99] Ryan C. Hendrickson, "NATO's Visegrad Allies: The First Test in Kosovo," *Journal of Slavic Military Studies*, Vol. 13, No. 2, 2000, pp. 30–31.

[100] IHS Janes, "External Affairs," *Jane's Sentinel Security Assessment*, August 14, 2014.

[101] "Full Text: John Boehner Speaks at the Unveiling of Havel Bust," *Prague Post*, November 19, 2014.

and Berlusconi, Zeman also might benefit from indirect, somewhat murky, financial ties to Russia. According to media accounts, two of the key financiers of Zeman's Strana Práv Občanů Party who are now among his inner circle of advisers are Martin Nejedly and Miroslav Slouf. Both also reportedly have extensive ties to the Russian oil and gas company Lukoil—with the former serving as the head of its Czech subsidiary and the latter helping Lukoil secure contracts in neighboring Slovakia.[102]

Despite these anecdotes, three key caveats must be kept in mind in assessing personal corruption's viability as a strategic tool. First, although Schröder, Berlusconi, and Zeman are clearly pro-Russian and have personally benefited from their relationship with Moscow, it is less clear whether Russia paid them to become pro-Russian or whether they were pro-Russian to begin with and Russia rewarded them as a result. Ultimately, this distinction is critical because it is the difference between an actual, intentional Russian hostile measure and a convenient, mutually profitable alignment of views. Russia simply cementing its friendships with financial and ideological ties is arguably less threatening than if it proves its ability to turn politicians in its favor.

Second and along similar lines, both Schröder and Berlusconi are former politicians, and—at least in the latter's case—already embroiled in scandal. Although Russia could conceivably use these figures as agents of influence in the future, their value is less than that of current leaders. Of the three cases, Zeman is arguably the most troubling because he still serves as president. Even in Zeman's case, though, the influence is diminished because most foreign policy decisions in the Czech Republic are made by the prime minister. Moreover, the Czech Republic has less weight in determining European foreign policy than its larger peers (such as Germany or Italy), although it still wields veto power as a member of the EU and NATO.

Finally, and most importantly, although it is relatively easy to see how Russia could use corruption to accomplish any number of its objectives—particularly undermining NATO solidarity and encour-

[102] Jan Richter, "Miloš Zeman—Political Veteran Seeking to Crown His Career," *Radio Praha*, December 19, 2012.

aging Russian political support—thus far, comments like those from Schröder, Berlusconi, and Zeman have seemingly produced negligible effects in shifting German, Italian, and Czech policy toward Russia. All in all, although personal corruption might help Russia achieve its objectives in Western Europe on the margins, it is unlikely to prove decisive now or over the coming years.

Information Operations

Of all the Russian hostile measures currently employed against the West, Russian information operations have attracted some of the most attention.[103] Following the Russian cyberattacks and social media engagement leading up to the 2016 U.S. election,[104] there was significant concern in France and Germany that Russian actors would use similar tools to influence upcoming European elections. Indeed, during the April–May 2017 French presidential election, the campaign of Emmanuel Macron (who ended up winning) was the victim of a major hack, with some emails publicly released shortly before the vote. The head of the U.S. National Security Agency, ADM Michael Rogers, noted that he had warned his French colleagues that "we're seeing them penetrate some of your [i.e., French] infrastructure."[105] In the September 2017 German federal elections, the far-right AfD won 12.6 percent of the vote, making it the first far-right party to gain seats in parliament in decades, even as Angela Merkel won her fourth term. Social media research conducted after the vote noted significant activity by "right wing internet trolls," including fake accounts on major sites criticizing parties other than AfD, although there was no specific indication that these activities were tied to Russia.[106]

---

[103] RAND MSW symposium, Cambridge, United Kingdom, February 2, 2016.

[104] Office of the Director of National Intelligence, 2017.

[105] Andy Greenberg, "The NSA Confirms It: Russia Hacked French Election Infrastructure," *Wired*, May 9, 2018.

[106] Charles Hawley, "Merkel Re-Elected as Right Wing Enters Parliament," Spiegel Online International, September 24, 2017; Michael Schwirtz, "German Election Mystery: Why No Russian Meddling?" *New York Times*, September 21, 2017, "Far-Right Trolls Active on Social Media Before German Election: Research," *Deutsche Welle*, February 21, 2018.

In addition to Russian activity on social media, Russian news sites are sometimes cited as a hostile influence. RT, one of the most visible outlets, has production quality comparable to Western outlets.[107] Even far outside Eastern Europe and Russia's traditional sphere of influence, RT's rise has generated a good deal of concern. Writing in *Politico*, Columbia Journalism School graduate Casey Michel argues, "Backed with a budget approaching $450 million in 2014, RT now acts as the tip of the Kremlin's information warfare machine, an agglomeration that seeks to undermine both notions of journalism and faith in the workings of liberal democracy."[108] Russian information operations also attracted policymakers' attention. The Chairman of the U.S. House Foreign Affairs Committee, U.S. Rep. Edward R. Royce, stated, "It's remarkable to see the sophisticated media offense that Putin is conducting across Eastern Europe, Central Europe, the Middle East, and Latin America through Russia Today." Royce and the committee's ranking minority member, Eliot Engel, offered legislation to counter Russian media influence.[109]

The actual threat of RT and related platforms, however, might not be that great. According to an investigation by the *Daily Beast* using documents leaked from a former RIA Novosti official, RT's reach and influence might be considerably less than is often portrayed. RT claims a "reach of 700 million people across more than 100 countries," but the figure represents potential viewers, not people who watch the channel.[110] The *Daily Beast* found that similar figures—such as the claim to have around 7 million viewers across six European countries—were extrapolated from phone interviewees who acknowledged having watched RT but were not regular viewers.[111]

---

[107] RAND MSW symposium, Cambridge, United Kingdom, February 2, 2016.

[108] Casey Michel, "Putin's Magnificent Messaging Machine," *Politico*, August 26, 2015.

[109] Guy Taylor, "Russia Propaganda Machine Gains on U.S.," *Washington Times*, December 27, 2015.

[110] Katie Zavadski, "Putin's Propaganda TV Lies About Its Popularity," *Daily Beast*, September 17, 2015.

[111] Zavadski, 2015.

Actual viewership data paints a different picture of RT's influence. The *Daily Beast* suggested that RT actual viewership might not constitute even 0.1 percent of Europe's television audience.[112] Other studies suggest RT's viewership might be slightly higher than that but still not significant. A study done in 2013 for the European Commission found that fewer than 0.5 percent of European audiences watched RT daily.[113] In contrast, more than 6 percent reported watching Sky News Daily, more than 4 percent said CNN, and more than 3 percent said BBC.[114] Monthly viewership figures paint an even starker picture. Fewer than 3 percent of those surveyed reported watching RT over the past month, compared with more than 25 percent for Sky News and BBC and more than 35 percent for CNN.[115] Recent evidence suggests that these statistics hold true today. Even in the United Kingdom, which reported some of the highest levels of RT viewership in 2013, 2016 data still showed a relatively small market share.[116] According to the Broadcasters' Audience Research Board, an estimated 688,000 watched RT during the week of March 6–13, 2016, reflecting an estimated 1.16 percent of the viewing audience, with the average amount of time spent watching amounting to just one minute.[117]

Moreover, not all RT content is the same. According to the *Daily Beast*'s survey of the RT's 100 most popular YouTube videos from 2010 to 2015, "natural disasters, accidents, crime, and natural phenomenon" clips attracted 81 percent (344 million views) of RT's viewership while political clips attracted 1 percent (fewer than 4 million views). Indeed, the study found that Putin's most popular video was not political at all

---

[112] Zavadski, 2015.

[113] Deirdre Kevin, Francesca Pellicanò, and Agnes Schneeberger, "Television News Channels in Europe," *European Audiovisual Observatory*, October 2013, p. 45.

[114] Kevin, Pellicanò, and Schneeberger, 2013, p. 45.

[115] Kevin, Pellicanò, and Schneeberger, 2013, p. 46.

[116] Zavadski, 2015.

[117] Broadcasters' Audience Research Board, "Weekly Viewing Summary," undated.

and featured him singing "Blueberry Hill" at a charity benefit in St. Petersburg in 2010.[118]

RT also runs up substantial costs. The Russian government spent 61.6 billion rubles—about $2 billion—on RT from 2005 to 2013.[119] By comparison, RT in 2013 reportedly cost twice as much as al Jazeera and 18 times more than Euronews—yet al Jazeera is one of the most popular channels in 32 countries and Euronews can make the same claim in 12 countries, whereas RT makes that list in only one. In fairness, the BBC cost about 30 percent more than RT—but BBC also considerably outperformed RT, finishing in the top ratings ranks in almost all markets considered.[120]

Aside from RT, Russia tried to influence Western audiences in other ways. In one lack-luster example, the Russian think tank Institute of Democracy and Cooperation recently opened offices in Paris and New York. Led by former State Duma deputy Natalya Narochnitskaya and funded by anonymous donors, the think tank promotes European-Russian relations and says that a "political order should be underpinned by a moral perspective, and specifically by the Judeo-Christian ethic which unites both the Eastern and Western parts of the European continent," a hat-tip to Russia's socially conservative agenda.[121] The Paris office of the Institute allegedly coordinates with the far-right National Front, although it denies this claim.[122]

Since opening its doors, the Institute of Democracy and Cooperation has proven less than successful. Its New York director, Andranik Migranyan, testified before Congress on Chechnya, for example, and for a time wrote for the left-leaning *Huffington Post* and the politi-

---

[118] Zavadski, 2015.

[119] Zavadski, 2015.

[120] Zavadski, 2015.

[121] Natalya Kanevskaya, "How the Kremlin Wields Its Soft Power in France," Radio Free Europe/Radio Liberty. June 24, 2014; Institute of Democracy and Cooperation, "The Institute of Democracy and Cooperation," undated.

[122] Kanevskaya, 2014.

cal realist National Interest.[123] In 2014, however, Migranyan made headlines for attacking Russian philosophy professor Andrei Zubov, who compared Putin's annexation of Crimea with Hitler's invasion of Austria and then Czechoslovakia's Sudetenland. Migranyan labeled Zubov "hell-spawn" and argued, "One should distinguish the difference between Hitler before 1939 and Hitler after 1939 and separate chaff from grain," suggesting that if Hitler only annexed Austria and Sudetenland, "he would have gone down in the history of his country as a politician of the highest order."[124] The Institute of Democracy and Cooperation in New York closed in June 2015, shortly after these comments were made. Migranyan claimed the think tank "accomplished its mission" and that "the human rights situation has improved in the United States."[125] Media accounts suggest the think tank faced financial constraints.[126]

By comparison, the European arm of Institute of Democracy and Cooperation is more successful. Director Narochnitskaya appears regularly on international media outlets, and director of studies John Laughland writes regularly for publications ranging from the British *Guardian* and *Spectator* to the *Hungarian Review*; the Briton also sits on the academic board of the Ron Paul Institute for Peace and Prosperity.[127] Laughland articulates the think tank's pro-Russian views, even promoting the idea that the EU was a plot by the Central Intelligence Agency to undermine Europe.[128]

---

[123] Andranik Migranyan, "Testimony on Russian-American Relations on the Question of Chechnya Before the U.S. Congress," House Committee on Foreign Affairs, Subcommittee on Europe, Eurasia, and Emerging Threats, April 26, 2013; Rosie Gray, "Pro-Putin Think Tank Based in New York Shuts Down," BuzzFeed, June 30, 2015.

[124] Neil MacFarquhar, "Russia Revisits Its History to Nail Down Its Future," *New York Times*, May 11, 2014.

[125] Gray. 2015.

[126] Gray, 2015.

[127] "John Laughland," *Hungarian Review*, undated; "Russia: Old Foe or New Ally?" Al Jazeera, December 12, 2015; Ron Paul Institute for Peace and Prosperity, "About the Institute," undated.

[128] Anne Applebaum, "Authoritarianism's Fellow Travelers," *Slate*, October 16, 2015.

In addition to the Institute for the Study of Democracy and Cooperation, Russia reportedly funds a variety of other think tanks, including Poland's European Center of Geopolitical Analysis, Estonia's Legal Information for Human Rights, Latvia's Institute of European Studies and its Human Rights Committee, Austria's World Public Forum Dialogue of Civilizations, and Serbia's Nasa Srbija.[129] The European Center of Geopolitical Analysis denies receiving Russian funding; others acknowledge Russian ties but deny that Russia directs their analysis.[130]

Russia also employs other means to spread its message, although the extent of these efforts varies widely by country. In Slovakia, for example, Russia actively offers scholarships to Slovaks to study in Russia, sponsors cultural goodwill tours, and uses internet trolls to help shape discussion. These efforts promote pan-Slavism while encouraging anti-Americanism.[131] In contrast, in Slovakia's Visegrad neighbors, Russian information operations finds a less receptive environment because the population is actively hostile (in Poland), indifferent (in the Czech Republic), or simply encounters language barriers (in Hungary).[132]

Ultimately, Russia's return on its investment is unclear. According to a Pew survey conducted between March and May 2015, Russia held unfavorability ratings of 66 percent in the United Kingdom and Spain, 67 percent in the United States, 69 percent in Italy, 70 percent in Germany and France, and 80 percent in Poland.[133] The percentage of respondents who had little or no confidence that Putin would "do the right thing regarding world affairs" proved similarly unfavorable: 76 percent in Germany, 77 percent in Italy, 80 percent in the United

---

[129] Elisabeth Braw, "The Kremlin's Influence Game," *World Affairs*, March 10, 2015b.

[130] Braw, 2015b.

[131] Interview with a former senior Slovak defense official, Bratislava, February 1, 2016.

[132] Interview with a former senior Czech defense official, Prague, February 1, 2016; interview with two think tank analysts in Cambridge, United Kingdom, February 2, 2016, and February 3, 2016.

[133] Bruce Stokes, *Russia, Putin Held in Low Regard Around the World: Russia's Image Trails the US Across All Regions*, Pew Research Center, August 5, 2015, p. 2.

Kingdom, 85 percent in France, 87 percent in Poland and 92 percent in Spain—although this percentage in the United States was only 75.[134] Moreover, the trends for both questions for most of these countries show little—or, in some cases, negative—progress; overall, Russia's image lags behind that of the United States.[135] Most importantly, high numbers of respondents stated that they consider Russia a threat: 91 percent of French, 89 percent of Polish, 89 percent of British, 88 percent of Italian, 87 percent of Spanish and 86 percent of German. In addition, 70 percent of Polish, 53 percent of British, 51 percent of French, and 49 percent of Spanish respondents labeled Russia a major threat.[136]

Some opinion surveys, of course, are more in Russia's favor. A Pew survey published in July 2015 asking about using military force against Russia if it attacked a NATO ally registered opposition from 58 percent of German, 53 percent of French, 51 percent of Italian, and 47 percent of Spanish respondents.[137] In Italy, 44 percent of respondents considered Russia a major threat; 38 percent of the German public thought so.[138] Notably, the Pew surveys did not cover some Southern and Central European countries that could be more pro-Russian, and it is possible that the polling figures would be even worse for Russia had it not been for Russian information operations. Still, if Russia wanted to shape public opinion in the West through information operations, the evidence suggests their rewards to date have been rather paltry. And so, although Russia will likely use information operations in the coming years—be it through RT or other means—the efficacy of these hostile measures remains an open question.

[134] Stokes, 2015, p. 5.

[135] Stokes, 2015, pp. 11, 15; Katie Simmons, Bruce Stokes, and Jacob Poushter, "NATO Publics Blame Russia for Ukrainian Crisis, but Reluctant to Provide Military Aid," Pew Research Center, June 10, 2015.

[136] Stokes, 2015, pp. 11, 15; Simmons, Stokes, and Poushter, 2015.

[137] Simmons, Stokes, and Poushter, 2015.

[138] Stokes 2015, pp. 11, 15; Simmons, Stokes, and Poushter, 2015.

Outreach to Russian Expatriates Abroad

Large ethnic Russian populations living abroad in Western Europe might provide Russia with another opportunity either to pressure these governments to adopt a more pro-Russian stance or destabilize these countries in the event of crisis. Accurate numbers of Russians living in Western Europe are relatively hard to come by, but many countries already have sizable ethnic Russian communities. Germany, for example, was home to 1,188,000 Russians in 2014.[139] Similarly, *The Guardian* placed the number of Russian expatriates living in London at about 150,000.[140] Moreover, the number of visas issued by the United Kingdom to Russians increased by 60 percent from 2010 to 2013.[141] Indeed, special visa laws specifically allowed Russians to gain permanent residency in the United Kingdom more quickly if they invested £1 million (pounds), £5 million, or £10 million in the United Kingdom.[142] In France, the Russian immigrant population makes up a relatively small amount of the total—a mere 2 percent of the approximately 200,000 immigrants to France 2012,[143] although some estimates place the total number of Russians living in France at anywhere between 200,000 and 500,000.[144]

Russia has attempted to maintain its connections to Russian expatriates in the West. In the immediate post–Cold War period, this outreach largely has been in terms of increasing cultural or economic ties.[145] According to some experts, these forms of outreach double as attempts to push these populations into advocating neutrality in any

---

[139] Destatis Statistiches Bundesamt, "Persons with a Migrant Background," undated.

[140] Viv Groskop, "How the Ukraine Crisis Is Affecting Russians in Moscow-on-Thames," *The Guardian*, April 6, 2014.

[141] Helen Warrell, "UK Sees Surge in Wealthy Russians with Fast-Track Entries," *Financial Times*, July 31, 2014.

[142] Warrell, 2014.

[143] Chantal Brutel, "Les Immigrés Récemment Arrivés en France [Immigrants Recently Arrived in France]," *Institut National de la Statistique et des Études Économiques*, No. 1524, November 2014.

[144] Demidoff, Maureen, "La Communauté Russe en France est «Éclectique [Russian Community in France Is Eclectic]," *Russie Info*, October 30, 2014.

[145] Steven Eke, "Russia President Targets Diaspora," BBC, October 24, 2006.

conflict with the West, if not into outright pro-Russian stances.[146] Media reports have also hinted at the existence of extensive Russian intelligence networks.[147]

Perhaps the best recent demonstration of how Russia can use these groups to its advantage comes from Germany. In January 2016, thousands of Russians living in Germany took to the streets protesting the alleged rape of a 13-year-old Russian-German girl by Muslim immigrants, while Russian Foreign Minister Lavrov accused German authorities of covering up the attack.[148] The German newspaper *Der Spiegel* even ran an extensive investigative report on the incident titled, "The Hybrid War: Russia's Propaganda Campaign Against Germany," and accused Russia of stirring up popular outrage about the rape—which German authorities insist never occurred—using the German broadcasts of the Russian media outlets of Sputnik and RT.[149] Ultimately, the case demonstrates a capability that Russia could exploit again if it ever wanted to create instability in a major Western capital.

But Russia would likely face a series of obstacles in attempting to use these expatriate communities for something larger. First, anecdotal accounts suggest that Russian immigration to the West is fueled by a desire to escape Russia or find a home for investments outside the reach of the Kremlin.[150] Indeed, United Kingdom laws encourage Russian capital flows to that country, and France traditionally has been one of the top destinations for Russian asylum seekers.[151] As a result, Russian diaspora communities in Western countries have personal stakes in ensuring that these countries remain stable and pros-

---

[146] Interview with a former senior Slovak defense official, Bratislava, February 1, 2016.

[147] For example, see Luke Harding, "Gordievsky: Russia Has as Many Spies in Britain Now as the USSR Ever Did," *The Guardian*, March 11, 2013; Bill Gertz, "Spy Ring Arrest Highlights Jump in Russian Spying Under Putin," *Washington Free Beacon*, January 28, 2015.

[148] Damien McGuinness, "Russia Steps into Berlin 'Rape' Storm Claiming German Cover-Up," BBC News, January 27, 2016a.

[149] Amann et al., 2016.

[150] Groskop, 2014.

[151] United Nations High Commissioner on Refugees, *Asylum Trends in 2014: Levels and Trends in Industrialized Countries*, Geneva, Switzerland, 2014, p. 11.

perous. Moreover, some evidence suggests that Russia might see its extensive diaspora community as both an asset and a strategic liability. In 2014, Russia passed a law forcing citizens living there to declare any foreign passports. The measure, sponsored State Duma Deputy Andrei Lugovoi (who is wanted by authorities in the United Kingdom in relation to the 2006 death of Russian security-services officer Aleksandr Litvinenko), stemmed from fears that Russians with ties to Western countries might present a security threat to the state.[152] Although the law did not apply to Russians living abroad, it underscores Russian fears about dual loyalty. Ultimately, although the Russian diaspora community might prove useful to the Kremlin for small-scale actions, Russia might face serious constraints in mobilizing the Russian diaspora on any large scale.

Shows of Force and Military Threats

Finally, Russia also seems to be using military exercises and the aggressive positioning of military assets to influence the West. Russia regularly conducts scheduled large-scale exercises, as well as "snap" exercises done on short notice. From 2013 to 2015, Russia has conducted 18 large-scale snap exercises, some with as many as 100,000 soldiers.[153] Although these exercises serve a variety of purposes, from increasing military readiness to providing cover for the annexation of Crimea, NATO Secretary General Jens Stoltenberg argues that they are also designed to "menace Russia's neighbors."[154]

Russia also routinely tests NATO allies' airspace. According to a 2015 NATO estimate, Russian air activity near (and in some cases, in) NATO allies' air space had increased by 70 percent since 2013, forcing NATO to intercept Russian aircraft some 400 times in 2015.[155]

---

[152] Carl Schreck, "Russian Expats Wrestle with Dual-Citizenship Dilemma," Radio Free Europe/Radio Liberty, March 14, 2014.

[153] Jens Stoltenberg, *The Secretary General's Annual Report:2015*, Brussels: NATO, 2016, p. 19.

[154] Stoltenberg, 2016, p. 19.

[155] Stoltenberg, 2016, p. 56.

Norway alone intercepted Russian warplanes 74 times in 2014.[156] Russian TU-95 Bear strategic bomber flights near the United Kingdom became an almost monthly occurrence in 2015. The flights always stayed outside Britain's sovereign airspace, and the British Ministry of Defence regarded them "more of a routine nuisance than a threat."[157] Russia even simulated a nuclear attack on the Swedish island of Gotland in 2013, 100 miles south of Stockholm, a move that reportedly caught the Swedish military off guard.[158]

Russia also has become more aggressive at sea. According to the Russian Navy chief Admiral Viktor Chirkov, Russia increased submarine patrols by 50 percent after 2013, including in and around Europe.[159] From October 2014 through April 2015, for example, there were at least three sightings of Russian submarines off the coast of Scotland and another two more instances near Finland and Sweden, respectively. Additionally, in April 2015, three Russian warships transited the English Channel.[160] There are also concerns that Russia could cut the transatlantic fiber-optic cables that carry much of U.S.-Europe communications, valued at an estimated $10 trillion a day in global business.[161] So far, aside from two suspected instances of submarines getting entangled with fishing trawler nets, these patrols have caused alarm but little physical damage.[162]

Extrapolating from the current trend lines, Russia will continue to conduct snap exercises, air flights, and maritime patrols but could

---

[156] Russell Goldman, "Russian Violations of Airspace Seen as Unwelcome Test by the West," *International New York Times*, October 6, 2015.

[157] Roland Oliphant, "Mapped: Just How Many Incursions into NATO airspace Has Russian Military Made?" *The Telegraph*, May 15, 2015.

[158] Roland Oliphant, "Russia 'Simulated a Nuclear Strike' Against Sweden, NATO Admits," *The Telegraph*, February 4, 2016.

[159] Demetri Sevastopulo, "Russian Navy Presents the US with a Fresh Challenge," *Financial Times*, November 2, 2015.

[160] Oliphant, 2015.

[161] David E. Sanger and Eric Schmitt, "Russian Ships Near Data Cables Are Too Close for U.S. Comfort," *International New York Times*, October 25, 2015.

[162] Oliphant, 2015.

confront logistical and strategic limitations. Logistically, Russia faces increasing costs for maintaining an aging fleet: The mainstay of Russia's strategic bomber force is the aging, prop-driven Tu-95, which first entered service in the 1950s and which the Russians historically struggled to keep flying.[163] Strategically, there are questions about what these shows of force accomplish: They are occasionally reported in the popular press and prompt NATO to scramble aircraft in response, but most of the concerns expressed by the United States and NATO focus on air safety and the risk of inadvertent escalation.[164] To date, if anything, these acts have led to more-robust U.S. and NATO presences, although Russia might retain a hope that it can intimidate the West in the future or confound the West's ability to differentiate between a real strike and a practice exercise.[165]

The Future of Russian Hostile Measures in Other Parts of Europe

Over the next few years, Russia will have the motivation, opportunity, and means to employ hostile measures against Western and Central Europe, but it is unclear just how successful it will ultimately be at accomplishing its aims. Indeed, the preceding analysis suggests a mixed finding. Although Russian efforts at hostile measures in Western and Central Europe might be able to capitalize on the host of challenges confronting Europe, Russia might also face a mismatch of motivation and opportunity and encounter significant obstacles in employing many of the tools it might use for hostile measures.

---

[163] Fred Weir, "Russia's Flights Over Europe: How Much Bark, How Much Bite?" *Christian Science Monitor*, October 30, 2014.

[164] Ryan Browne and Jim Sciutto, "Russian Jets Keep Buzzing U.S. Ships and Planes. What Can the U.S. Do?" CNN, April 19, 2016.

[165] Browne and Sciutto, 2016.

Multiple Exploitable Economic and Political Multiple Cleavages

In terms of the macropicture, Europe faces a series of crises on several fronts—immigration, terrorism, economics, and dissatisfaction with the EU—that increase its vulnerability to hostile measures now and over the short to medium term. Being mired in these other problems can weaken Europe's resolve to aggressively confront Moscow. These conditions foster the rise extremist parties—on both the left and the right—that will take a softer line on Russia. More importantly, these crises create natural regional fissures in Europe that Russia can further exacerbate using information operations and proxies. Importantly, none of these crises were created by Russia but do present Russia with strategic opportunities that it can exploit with hostile measures to chip away at European unity and frustrate U.S. strategic aims.

Potential Motivation and Opportunity Mismatches

At the state level, Russia increasingly might find itself confronting mismatches in motivation and opportunity over the next few years. From a Russian strategic perspective, the larger countries of Western Europe present a more attractive target. They retain the bulk of the political, economic, and military power both in Europe and worldwide. If Russia could persuade any of these countries to take a more pro-Russian line—or at least a more anti-U.S. one—it would present a major strategic success for Russia and deal the United States a significant blow. Even if Russia could sow disunity with only the EU and/or NATO, it could still be considered a win.

But just because Russia might want to influence these countries does not necessarily mean that it will have the opportunity to do so. Some of these countries (for example, Germany and the United Kingdom) have done comparatively well economically and are thus well positioned to resist efforts by Russia and its proxies. Others—such as France—have managed to avoid the brunt of the immigration crisis and have smaller Russian expatriate populations. From an opportunity perspective, Russia might have more success in those countries harder hit by the economic crisis, the immigration crisis (i.e., Southeastern Europe), or where there are larger Russian expatriate and Slavic populations (i.e., Central and Eastern Europe). Although Western Europe is not immune to Russian hostile measures—far from it—there are bounds on what Russia can and cannot accomplish through these tools.

All Hostile Measures Tools Are Not Equally Valuable and Some Are in Decline

Russia's capacity to conduct hostile measures is not boundless, either. Although Russia stands to reap the benefits of its association with both left- and right-wing extremist parties over the next few years as the influence of these parties increases, these parties might have grown because they have inherent constituencies in many European countries, not because of direct Russian intervention. Russia simply had to provide a little help when these parties needed it. Moreover, it remains to be seen how much influence, if any, Russia can exert over these parties over the longer term. These parties are ultranationalists, not simply Russian puppets. The extreme right and left might share some of Russia's interests for the moment, but that does not mean they are bound to follow in lockstep going forward.

Russia will likely also employ other hostile measures in the coming years, but it is less clear how effective they will be. Some—like corruption, information operations, outreach to Russian exiles and Russian shows of military force—are attention-grabbing but likely less of a threat to U.S. national strategy, at least in the short term. It is possible to document how Russia employs all three tools and imagine how it might do so in the future, but proving that these tools meaningfully altered European or NATO policy is more difficult. Finally, Russia's economic leverage, particularly with European energy supplies, was considerable but is diminishing as Europe gains alternative energy sources and Russia faces its own economic difficulties.

Ultimately, Russian hostile measures directed against Europe should remain a concern but also should be viewed in perspective. Russia will enjoy significant opportunities to influence these parts of Europe in the short term but will encounter significant obstacles in trying to use hostile measures. Russia's influence through hostile measures might not be 12 feet tall, but it is not a dwarf, either.

CHAPTER FIVE

Conclusions

Russia has a wide range of tools and methods short of conventional war that it can use in efforts to achieve its political, military, and economic goals in Europe. Although there is no way to predict what Russia will do, it is possible to analyze Russia's motives and opportunities, as well as the potential means it might employ. There are significant methodological challenges because Russia's activities are covert or deniable by their very nature. As discussed in Chapter One, in an ideal world, this study would be able to document the intent of Russian leadership, relevant Russian proxies in Europe, and how the activities of these proxies achieved Russian objectives. In practice, this information is only rarely available, especially through open sources. Even with these data limitations, however, we can draw upon the insights of regional experts and the available literature to reach some basic conclusions about the size, scale, and methods of Russian hostile measures—and, as importantly, what the United States, the Joint Force and the U.S. Army should do about them.

The Nature of the Threat

Russian hostile measures come in a variety of forms—including economic pressure, information operations, political corruption, support for ethnic minorities and separatist movements, and more-limited and deniable uses of force (i.e., "little green men" scenarios). Based on analysts' accounts of overall Russian strategic perspective and observations about how Russia is pursuing its objectives across Europe, Russia appears to be adopting a "soft strategy" in which it seeks to achieve its overall objectives by applying a wide and flexible array of hostile measures across instruments of national power to generate possibilities and shape conditions. The threat of these tactics depends on the vulnerability of the different countries and the resources that Russia can deploy to achieve its tactics.

Fortunately, there are no obvious major vulnerabilities that Russia can easily exploit to its advantage within NATO. Indeed, there generally seems to be an inverse correlation between the likelihood of Russia employing tactics and those tactics' potential impacts on NATO and U.S. security. For example, although Russia currently produces propaganda, engages in targeted corruption, and uses economic pressure—and will likely continue to do all this for the foreseeable future, the impact of these tactics on European security and U.S. interests has been questionable at best. Conversely, the likelihood that Russia will use covert or denied military means—little green men—in a NATO country appears quite low, although such activity, if it were adopted, could pose a significant risk of escalating to full-scale conflict. In other words, in assessing the risk of Russian hostile measures over the next few years, there seem to be trade-offs between probability and risk.

With respect to identifying the regions within the EU and NATO that face the greatest threat, understood as the combination of Russia's capabilities and intent, it appears that the Baltics and Southeastern Europe are more threatened than other parts of Europe for a variety of economic, cultural, historic, and governance reasons. The Baltic countries, with stronger rule of law and significant effort to integrate Russian-speaking populations, are relatively resistant to Russian subversion but much closer geographically to Russia and more isolated from the rest of the EU and NATO. Bulgaria, Greece, and Romania have fewer government resources at their command and continuing economic and political challenges, and thus less capacity to respond. However, except for Bulgaria and Greece, these countries face less direct pressure from Russia. The danger of Russian subversion within these countries is largely contained to the region. Although these countries are members of the EU and NATO and do have veto power over the policies of these organizations, their economic and security dependence on these organizations means that they are unlikely to unilaterally challenge consensus views.

Russia could undermine the EU and NATO more directly if it could affect the major Western European countries, including the United Kingdom, France, Germany, and Italy. To date, Russian hostile measures do not appear to pose such a risk. Instead, the greatest threat might be internal—the United Kingdom's planned exit from the EU, for example, has the potential to weaken the EU significantly. Russia might attempt to take advantage of the internal threats facing EU countries, such as migration and economic dissatisfaction, to further divide both EU and NATO, but it will face an uphill battle, based on the analysis above.

The United States has expressed significant concern about Russian hostile measures. As the 2018 National Security Strategy notes,

> Russia is using subversive measures to weaken the credibility of America's commitment to Europe, undermine transatlantic unity, and weaken European institutions and governments. . . . The United States is safer when Europe is prosperous and stable, and can help defend our shared interests and ideals. The United States remains firmly committed to our European allies and partners.[1]

This study, and others like it, have focused primarily on Russian activity in NATO countries because the United States has more strongly articulated interests and a greater role in defending and deterring aggression in these societies. Understandably, these countries are of primary concern to U.S. policymakers; the likelihood of U.S. military and political action in response to Russian aggression is far greater.

However, Russia has significant and perhaps greater ability to destabilize and undermine non-EU and non-NATO countries, such as Belarus, Bosnia and Herzegovina, Serbia, Moldova, and Ukraine. Although the 2018 National Security Strategy affirms U.S. commitment to supporting its partners, the level and scope of U.S. interest and commitment to these countries, and hence the prioritization and willingness of the United States to invest significant resources, remains under debate. It is notable that the U.S. military, for example,

---

[1] White House, *National Security Strategy of the United States of America*, December 2017, pp. 47–48.

is engaged in training, exercises, and support for Moldova, Ukraine, and Georgia, and could face a direct confrontation with Russian forces or their proxies engaged in these countries. Additional research on U.S. and Russian interests, operations, and the risk of conflict is needed to evaluate and improve U.S. foreign policy in non-EU and non-NATO countries, and to prepare for the greater risk of conflict that might emerge from within these societies.

The Contours of the Response

Russian hostile measures are undertaken by the whole of the Russian government and enlist NGOs, such as companies, oligarchs, religious organizations, and foundations. Consequently, a full and adequate U.S. response will require engaging the whole of the U.S. government, similarly supported by NGOs. For example, Russian cooption of criminal enterprises might require support for building the rule of law; Russian support for political parties might be countered by assistance to political parties from the National Democratic Institute and International Republican Institute; and countering Russian information operations might involve a robust engagement by the Broadcasting Board of Governors. The U.S. military has a key role to play in deterring aggression (and possibly in assisting responses), but, in many cases, it will not have a leading role. Our analysis does not include a detailed assessment of prioritization of resources or variation in levels of desired engagement based on uncertainty or debate about the nature of U.S. interests in non-EU or non-NATO countries. Because our focus is on countering Russian hostile measures, we conducted the analysis in Table 5.1 assuming a strong degree of U.S. interest throughout the region. In practice, the desired policy response might be different depending on the level of resources and engagement that the United States seeks.

Table 5.1 outlines the major hostile measures that Russia might adopt in Europe, identifying them based on whole-of-government (WoG) and military solutions.

Taken as whole, Table 5.1 emphasizes that much of the response to Russian hostile measures revolves around increasing political development across Europe, including building institutional resilience,

Table 5.1
Forecast of Russian Hostile Measures in Europe in the Next Few Years

Region	Measure	Likelihood	Severity of Impact	U.S. Options to Mitigate
Baltics	Ethnic conflict/ support for pro-Russian groups	High	Low	**High concern** • *WoG:* Diplomatic and economic support to resolve minority grievances • *Military:* Positive engagement with Russian-speaking communities when deployed
Baltics	Support for non-Russian ethnic groups	Low	Low	**Low concern** • *WoG:* Improved indicators and warnings; improved pro-Western strategic communication; support for investigative journalism • *Military:* Increased intelligence-gathering effort to understand and warn about these efforts
Baltics	Covert or deniable military action	Low	Moderate	**Moderate concern** • *WoG:* Indicators and warnings; support for Baltic government and security forces; support for Russian-speaking populations • *Military:* Increased presence to signal credible deterrent, better intelligence; civil affairs and related activities to increase trust of local Russian speakers; exercises with host-country forces to ensure readiness for a wide range of contingencies
Baltics	Economic leverage	Moderate	Low (decreasing dependence)	**Low concern** • *WoG:* Improved activities to track Russian financing; support for Baltic efforts to diversify energy supplies and trade • *Military:* N/A
Baltics	Bullying with military/ intelligence forces	High	Low	**Low concern** • *WoG:* Diplomatic reassurance to allies; possible negotiation with Russia to develop confidence-building measures, transparency, and other options to reduce risk of escalation • *Military:* U.S. exercises to respond; presence for deterrence and reassurance

Table 5.1—Continued

Region	Measure	Likelihood	Severity of Impact	U.S. Options to Mitigate
Western and Central Europe	Personal corruption	High	Low/ moderate	**Low/moderate concern** • *WoG:* Support for rule of law in Central Europe; U.S. efforts to undermine Russian organized crime • *Military:* N/A
Western and Central Europe	Political influence	High	Moderate	**Moderate concern** • *WoG:* Support for investigative journalism (follow the money); strategic communications • *Military:* N/A
Western and Central Europe	Shows of force/ military threats	High	Low/ moderate	**Low/moderate concern** • *WoG:* Diplomatic efforts to clarify rules of engagement and ensure crisis management • *Military:* U.S. exercises to respond; presence for deterrent and reassurance
Western and Central Europe	Economic leverage	Moderate	Low	**Low concern** • *WoG:* Support for rule of law, diversification of energy • *Military:* N/A
Western and Central Europe	Propaganda information operations	High	Low	**Low concern** • *WoG:* Support for investigative journalism efforts in uncovering and debunking Russian propaganda • *Military:* Coordination and awareness in developing messaging during Army operations; preparation and training prior to operations
Western and Central Europe	Outreach to Russian exiles abroad	Moderate	Low	**Low concern** • *WoG:* Study and identification of Russian networks throughout Europe; support for host-country efforts • *Military:* N/A

Table 5.1—Continued

Region	Measure	Likelihood	Severity of Impact	U.S. Options to Mitigate
Southeastern Europe	Economic influence/ oligarchs	High	Moderate	**Moderate concern** • *WoG:* Increased U.S. assistance for democracy, civil society, rule of law, economic development; support for anti-corruption measures • *Military:* N/A
Southeastern Europe	Information warfare	High	Low/ moderate	**Low/moderate concern** • *WoG:* Support for investigative journalist efforts in uncovering and debunking Russian propaganda • *Military:* Coordination and awareness when deployed; preparation and training prior to operations
Southeastern Europe	Political influence	High	Moderate	**Moderate concern** • *WoG:* Intelligence effort to expose links; institution-building and transparency initiatives to promote good governance • *Military:* N/A
Southeastern Europe	Support for ethnic/separatist conflict	High	Moderate	**Moderate concern** • *WoG:* Support for civil society, economic development; support for investigative journalism to uncover Russian involvement • *Military:* Continued presence in the Balkans to prevent reigniting frozen conflicts; preparation and exercises in case of renewed conflict
Southeastern Europe	Military influence	Low	Moderate	**Low concern** • *WoG:* Analysis and preparation for contingencies in the region; diplomatic communications about what the United States would and would not do • *Military:* Forward presence to deter Russian military actions; U.S. exercises to respond; presence for deterrent and reassurance

enhancing the rule of law, and developing democratic accountability.[2] Russian hostile measures capitalize on state weakness, and far greater U.S. effort to address such weakness might be necessary in countering Russian measures. In the case of EU member states, U.S. assistance significantly declined following EU accession, based in part on the assumption that the EU would take the lead in supporting political development in Europe. Although there have been notable successes of the EU in these countries, the vulnerabilities as outlined remain. Successful U.S. re-engagement will depend on a coordinated, WoG effort involving a range of U.S. government organizations, including the Department of State, the U.S. Agency for International Development, and the Department of Justice.

There might be at least two challenges in developing a major WoG re-engagement in EU and NATO countries. First, some countries, especially the Baltic states, might not be supportive of U.S.-led actions to increase resilience to Russian hostile measures. For example, the Baltic countries believe that they should have the lead role in developing policy toward Russian speakers and tend to downplay the concerns of Russian speakers in their country. It is unclear how to persuade Baltic countries of the need for U.S. assistance,[3] but tying greater U.S. military assistance to accepting outside aid might be feasible. Second, it is unclear how to create a more coordinated U.S. approach to improving political development in the region that responds to a strategic threat from Russia. For the most part, these tasks fall outside the military's lane and require other departments and agencies to take the lead. Although the United States often paid lip service to this idea during the Iraq and Afghanistan wars, in practice, the military often led the response, with mixed success.[4] U.S. government efforts are not consistently designed to address strategic priorities, partly based on

---

[2] See, for example, Francis Fukuyama, *The Origins of Political Order: From Prehuman Times to the French Revolution*, New York: Farrar, Straus, and Giroux, 2014.

[3] See Radin, 2017a.

[4] See Linda Robinson, Paul D. Miller, John Gordon IV, Jeffrey Decker, Michael Schwille, and Raphael S. Cohen, *Improving Strategic Competence: Lessons from 13 Years of War*, Santa Monica, Calif.: RAND Corporation, RR-816-A, 2014.

the premise that tying them to military and political priorities would undermine effectiveness of the latter two.[5] U.S. policymakers will need to reevaluate the relationships among and coordination of the various agencies to ensure that U.S. government efforts effectively address the full range of threats posed by Russia.

Implications for the Joint Force and the U.S. Army

Although DoD and the U.S. Army will have a lead role only in certain Russian hostile measures (usually those involving the direct threat of military force), the Army will still play an important supporting role for several reasons. In many countries and areas, U.S. military forces or personnel might be the most capable U.S. government organization present. In other cases, the Army has unique capabilities it can bring to bear. Finally, in some cases, there is a direct link between the U.S. Army's role in deterring conventional Russian aggression and responding to Russian hostile measures. We identify three critical lessons for the Army to best perform its various roles.

First, the U.S. military as a whole—and the U.S. Army in particular—should prepare any forward presence designed to address the Russian conventional threat to also defend and counter Russian hostile measures. RAND research has emphasized the conventional vulnerability of the Baltics, and there might be other conventional vulnerabilities elsewhere in Europe.[6] Forces deployed to ensure conventional deterrence can have both positive and negative impacts on the risk of hostile measures. For example, Russian speakers in Estonia have voiced their opposition to the deployment of NATO forces, indicating a risk of a local backlash against U.S. forces.[7] Through careful preparation, and attention to manning, training, and equipping, the

---

[5] See for example, Max Boot and Michael Mikaucic, "Reconfiguring USAID for State-Building," Council of Foreign Relations, Policy Innovation Memorandum No. 57, June 22, 2016.

[6] Johnson and Shlapak, 2015.

[7] Radin, 2017b.

U.S. military can ensure that its deployments better defend against a range of Russian hostile measures. This includes getting better intelligence and counterintelligence, providing improved indicators and warning against Russian military action, preventing frozen conflicts from reigniting, and building institutional capacity in partner-nation armed forces.

There are various opinions within NATO as to whether increased deployment increases the risk of "poking the bear"—that whatever military presence the United States or NATO places in contested areas proves sufficient to provoke but not deter Russian actions.[8] There is probably no definitive answer to what will provoke Russia as opposed to deterring it; Russian leaders themselves are likely still evaluating NATO's posture. Thus, the United States must be attentive to potential allied and Russian redlines, and remain aware that Russia will evaluate deployed U.S. forces based on their potential to achieve regime change. The United States should not base its security on Russian demands, but it can consider the risk of war from Russian miscalculation in deploying its forces.

Second, the U.S. military should seek to develop greater capacity and capabilities for certain enablers that are especially relevant for Russian hostile measures. Intelligence resources—particularly counterintelligence, advanced human intelligence and (in some cases) signals intelligence assets—are required to detect and identify the employment of certain Russian hostile measures. To a lesser extent, perhaps, these development measures will put a new focus on public affairs personnel (particularly to counter Russian information operations about U.S. military activities) and military information support operations personnel (to help shape the environment). Depending on the country, these measures also might provide a new mission for civil affairs personnel as they attempt to build institutional capacity and interact with the citizenry. Although the U.S. military might not have the lead role in these tasks, its personnel might have a key role in any such activities,

---

[8] Stephanie Pezard, Andrew Radin, Thomas Szayna, and F. Stephen Larrabee, *European Relations with Russia: Threat Perceptions, Responses and Strategies in the Wake of the Ukraine Crisis*, Santa Monica, Calif.: RAND Corporation, RR-1579-A, 2017.

and it is critical to ensure that sufficient personnel with the appropriate training and preparation are deployed to perform these tasks. Given the likely long-term adversarial relationship between Russia and the United States in terms of military engagement in Eastern Europe, one idea for the Army to consider is greater training and extended deployment timelines for Foreign Area, Public Affairs, Civil Affairs, and Information Warfare Officers and other relevant specialties working in Eastern European countries. If these personnel have greater expertise, spend more time in-country, and continue to work on the same issues for many years, they might be better able to contribute to the U.S. Army and DoD missions in these regions.

Finally, responding to Russian hostile measures places a new premium on political awareness, as well as on crisis management and response. This report's discussion of Russian strategy notes that Russia seeks to encourage tension and exploit opportunity when crises emerge. The U.S. military can counter these Russian actions in two ways. First, military personnel, especially those deployed in countries with frozen conflicts or where there is a large pro-Russian population, need to take an active role in preventing crises from emerging. When dealing with Russian shows of force or handling long-standing ethnic divisions tied to frozen conflicts, the U.S. military as a whole and the U.S. Army specifically might wind up handling a political powder keg. Soldiers at all levels need to be aware of the political sensitivities involved and the risks of international escalation and need to be prepared to act appropriately. Second, an effective and timely U.S. government response is critical when crises do emerge. This requires a preexisting understanding of the political, economic, and military situation in all countries in the region, along with the actions that Russia could take. Policymakers are reasonably focused on NATO and especially on the Baltics, but it is necessary to expand preparations to deal with a wide range of possible crises. The Russian interventions in Ukraine in 2014 and in Georgia in 2008 point to the significant risk of Russian aggression in another non-NATO country in the future. Whatever the U.S. response, preparation for involvement in a wide range of conflicts can help reduce the risk of mismanagement, miscalculation, and escalation.

References

"22% of Bulgarians Want to Join Russia's 'Eurasian Union,'" EurActiv.com, May 15, 2015. As of March 23, 2016:
https://www.euractiv.com/section/elections/news/22-of-bulgarians-want-to-join-russia-s-eurasian-union/

Abdelal, Rawi, "The Profits of Power: Commerce and Realpolitik in Eurasia," *Review of International Political Economy*, Vol. 20, No. 3, June 2013, pp. 421–456.

"Agriculture Minister: Russian Food Will Squeeze Out Imports in 10 Years," *Moscow Times*, July 7, 2015. As of April 2016:
http://www.themoscowtimes.com/business/article/agriculture-minister-russian-food-will-squeeze-out-imports-in-10-years/525236.html

Aleksashenko, Sergey, "For Ukraine, Moldova, and Georgia, Free Trade with Europe and Russia Is Possible," Carnegie Middle East Center, July 3, 2014. As of March 2016:
http://carnegie-mec.org/2014/07/03/for-ukraine-moldova-and-georgia-free-trade-with-europe-and-russia-is-possible

Amann, Melanie, Markus Becker, Benjamin Bidder, Hubert Gude, Konstantin von Hammerstein, Alexej Hock, Christiane Hoffmann, Martin Knobbe, Peter Maxwill, Peter Müller, Gordon Repinski, Sven Röbel, Anna Sadovnikova, Matthias Schepp, Jörg Schindler, and Christoph Schult, "The Hybrid War: Russia's Propaganda Campaign Against Germany," Spiegel Online International, February 5, 2016. As of March 15, 2016:
http://www.spiegel.de/international/europe/putin-wages-hybrid-war-on-germany-and-west-a-1075483.html

Andreev, Alexander, "Bulgaria: Caught Between Moscow and Brussels," *Deutsche Welle*, April 27, 2014. As of June 2016:
http://www.dw.com/en/bulgaria-caught-between-moscow-and-brussels/a-17594779

Antonenko, Oksana, "Russia and the Deadlock over Kosovo," *Survival*, Vol. 49, No. 3, 2007, pp. 91–106.

Applebaum, Anne, "Authoritarianism's Fellow Travelers," *Slate*, October 16, 2015. As of March 23, 2016:
http://www.slate.com/articles/news_and_politics/foreigners/2015/10/vladimir_putin_s_fellow_travelers_the_russian_president_has_an_assortment.html

Assenova, Margarita, "Bulgaria Quits Belene Nuclear Power Plant, Open Doors to South Stream," *Eurasia Daily Monitor*, Vol. 9 No. 65, April 2, 2012. As of July 18, 2016:
http://www.jamestown.org/programs/edm/single/?tx_ttnews%5Btt_news%5D=39216&cHash=75258f500b9e78207d2fe6088ddd0b31#.V4z0Xmf2aUk

Association of Tour Operators, "Serbia Saw a 25% Increase in Tourist Arrivals from Russia in 2012," March 20, 2013. As of July 18, 2016:
http://www.atorus.ru/en/news/press-centre/new/21400.html

Baker, Peter, and Steven Erlanger, "Russia Uses Money and Ideology to Fight Western Sanctions," *New York Times*, June 7, 2015. As of March 16, 2016:
http://www.nytimes.com/2015/06/08/world/europe/russia-fights-wests-ukraine-sanctions-with-aid-and-ideology.html

Balmforth, Tom, "Russians of Narva Not Seeking 'Liberation' by Moscow," Radio Free Europe/Radio Liberty, April 4, 2014. As of June 1, 2016:
http://www.rferl.org/content/russia-estonia-not-crimea/25321328.html

"Baltic States Will Build New NPP In Lithuania," *Baltic Review*, January 6, 2016. As of June 2, 2016:
http://baltic-review.com/baltic-states-npp/

Balzer, Harley, "The Ukraine Invasion and Public Opinion," *Georgetown Journal of International Affairs*, March 20, 2015. As of March 2016:
http://journal.georgetown.edu/spotlight-on-16-1-inequality-the-ukraine-invasion-and-public-opinion/

Banse, Dirk, Florian Flade, Uwe Müller, Eduard Steiner, and Daniel Wetzel, "Circles of Power: Putin's Secret Friendship with Ex-Stasi Officer," *The Guardian*, August 13, 2014. As of March 11, 2016:
http://www.theguardian.com/world/2014/aug/13/russia-putin-german-right-hand-man-matthias-warnig

Bartles, Charles K., and Roger N. McDermott, "Russia's Military Operation in Crimea," *Problems of Post-Communism*, Vol. 61, No. 6, 2015, pp. 46–63.

Batchelor, Tom, "The New Iron Curtains: Where the Fences Are Going Up Across Europe to Keep Migrants Out," *Express*, December 6, 2015. As of March 21, 2016:
http://www.express.co.uk/news/world/624488/Europe-border-fences-migrant-crisis

Bechev, Dimitar, *Russia in the Balkans: How Should the EU Respond?* Brussels: European Policy Centre, October 12, 2015. As of July 15, 2016:
http://www.epc.eu/documents/uploads/pub_6018_russia_in_the_balkans.pdf

———, *Russia's Influence in Bulgaria*, Brussels: New Direction, The Foundation for European Reform, February 24, 2016. As of July 14, 2016:
http://europeanreform.org/index.php/site/publications-article/russias-influence-in-bulgaria

———, *Rival Power: Russia's Influence in Southeast Europe*, New Haven, Conn.: Yale University Press, 2017.

Belous, Sergey, "How Long Will Belgrade Seesaw Between NATO and Russia?" *Oriental Review*, April 23, 2016. As of July 2016:
http://orientalreview.org/2016/04/23/how-long-will-belgrade-seesaw-between-nato-and-russia/

Berzina, Ieva, "Zapad 2013 as a Form of Strategic Communication," in Liudas Zdavavičius and Matthew Czekaj, eds., *Russia's Zapad 2013 Military Exercise: Lessons for Baltic Regional Security*, Washington, D.C.: Jamestown Foundation, 2015. As of April 2016:
https://jamestown.org/wp-content/uploads/2015/12/Zapad_2013_-_Full_online_final.pdf

Beta, "Serbian Parties Sign Declaration with United Russia," B92, June 29, 2016. As of July 18, 2016:
http://www.b92.net/eng/news/politics.php?yyyy=2016&mm=06&dd=29&nav_id=98469

Bieber, Florian, *Post-War Bosnia: Ethnicity, Inequality and Public Sector Governance*, New York: Palgrave Macmillan, 2006.

Bigg, Claire, "Ukraine Livid as Putin, Berlusconi Swig Crimea's Oldest Bottle of Wine," Radio Free Europe/Radio Liberty, September 17, 2015. As of March 11, 2016:
http://www.rferl.org/content/ukraine-crimea-putin-berlusconi-drank-oldest-bottle-of-wine/27254178.html

Bildt, Carl, *Russia, the European Union, and the Eastern Partnership*, Latvia: European Council on Foreign Relations, ECFR Riga Series, May 19, 2015. As of March 2016:
http://www.ecfr.eu/page/-/Riga_papers_Carl_Bildt.pdf

Birnbaum, Michael, "Gay Rights in Eastern Europe: A New Battleground for Russia and the West," *Washington Post*, July 25, 2015. As of April 2016:
https://www.washingtonpost.com/world/europe/gay-rights-in-eastern-europe-a-new-battleground-for-russia-and-the-west/2015/07/24/8ad04d4e-2ff2-11e5-a879-213078d03dd3_story.html

Blank, Stephen, "Putin Sets His Eyes on the Balkans," *Newsweek*, April 17, 2015. As of July 15, 2016:
http://www.newsweek.com/putin-sets-his-eyes-balkans-323002

———, "Russia's Newest Balkan Games," *Eurasia Daily Monitor*, Vol. 13, No. 47, March 9, 2016. As of July 18, 2016:
http://www.jamestown.org/single/?tx_ttnews%5Btt_news%5D=45186&no_cache=1#.V40Ojmf2aUl

Blome, Nikolaus, Susanne Koelbl, Peter Müller, Ralf Neukirch, Matthias Schepp, and Gerald Traufetter, "Putin's Reach: Merkel Concerned About Russian Influence in the Balkans," Spiegel Online International, November 17, 2014. As of July 18, 2016:
http://www.spiegel.de/international/europe/germany-worried-about-russian-influence-in-the-balkans-a-1003427.html

Boot, Max, and Michael Mikaucic, "Reconfiguring USAID for State-Building," Council of Foreign Relations, Policy Innovation Memorandum No. 57, June 22, 2016. As of July 2016:
http://www.cfr.org/foreign-aid/reconfiguring-usaid-state-building/p37869

Borger, Julian, "Vladimir Putin Moves to Strengthen Ties with Serbia at Military Parade," *The Guardian*, October 16, 2014. As of July 2016:
https://www.theguardian.com/world/2014/oct/16/vladimir-putin-russia-serbia-alliance-military-parade

Braw, Elisabeth, "Russian Spies Return to Europe in 'New Cold War,'" *Newsweek*, December 10, 2014. As of March 21, 2016:
http://www.newsweek.com/2014/12/19/spies-are-back-espionage-booming-new-cold-war-290686.html

———, "Putin Seeks to Influence Radical Parties in Bid to Destabilise Europe," *Newsweek*, January 9, 2015a. As of March 16, 2016:
http://www.newsweek.com/2015/01/16/putins-envoys-seek-influence-european-radicals-297769.html

———, "The Kremlin's Influence Game," *World Affairs*, March 10, 2015b. As of March 23, 2016:
http://www.worldaffairsjournal.org/blog/elisabeth-braw/kremlin%E2%80%99s-influence-game

———, "Bully in the Baltics: The Kremlin's Provocations," *World Affairs Journal*, March/April 2015c. As of April 2016:
http://www.worldaffairsjournal.org/article/bully-baltics-kremlin's-provocations

Broadcasters' Audience Research Board, "Weekly Viewing Summary," undated. As of March 23, 2016:
http://www.barb.co.uk/viewing-data/weekly-viewing-summary/

Browne, Ryan, and Jim Sciutto, "Russian Jets Keep Buzzing U.S. Ships and Planes. What Can the U.S. Do?" CNN, April 19, 2016.

Brüggemann, Karsten, and Andres Kasekamp, "The Politics of History and the 'War of Monuments' in Estonia," *Nationalities Papers*, Vol. 36, No. 3, 2008.

Brunwasser, Matthew, and Dan Bilefsky, "After Protests, Prime Minister Resigns," *New York Times*, February 20, 2013. As of July 18, 2016:
http://www.nytimes.com/2013/02/21/world/europe/bulgarian-government-is-reported-set-to-resign.html

Brutel, Chantal, "Les Immigrés Récemment Arrivés en France [Immigrants Recently Arrived in France]," *Institut National de la Statistique et des Études Économiques*, No. 1524, November 2014. As March 15, 2016:
http://www.insee.fr/fr/themes/document.asp?reg_id=0&ref_id=ip1524

"Bulgaria, Greece in Action to Lure Russian Tourists," *Hurriyet Daily News*, March 20, 2016. As of July 18, 2016:
http://www.hurriyetdailynews.com/bulgaria-greece-in-action-to-lure-russian-tourists.aspx?pageID=238&nID=96676&NewsCatID=349

Bullough, Oliver, "Former Aide Says Putin Has No Strategic Plans," *Time*, November 5, 2014. As of March 2016:
http://time.com/3547935/putin-pugachev-oligarchs/

Burgen, Stephen, "Podemos Leaders Deny Venezuela Government Funding Link Claims," *The Guardian*, April 6, 2016. As of October 25, 2016:
https://www.theguardian.com/world/2016/apr/06/podemos-spain-leaders-deny-funding-link-venezuela-chavez-government

Cage, Sam, and Tsvetelia Tsolova, "Bulgarian Government Resigns Amid Growing Protests," Reuters, February 20, 2013. As of July 19, 2016:
http://www.reuters.com/article/us-bulgaria-government-idUSBRE91J09J20130220

Carlin, John, "Pablo Iglesias: "Catalans, Stay with Us and Let's All Kick Out Rajoy Together," *El País*, September 24, 2015. As of March 22, 2016:
http://elpais.com/elpais/2015/09/24/inenglish/1443088078_061646.html

Center for International Development and Conflict Management, "Assessment for Magyars (Hungarians) in Romania," College Park, Md.: Minorities at Risk Program, University of Maryland, undated. As of July 2016:
http://www.mar.umd.edu/assessment.asp?groupId=36002

Center for the Study of Democracy, *Energy Sector Governance and Energy (In) security in Bulgaria*, Sofia, 2014.

Central Intelligence Agency, "GDP (Purchasing Power Parity)," *World Factbook*, undated. As of March 31, 2016:
https://www.cia.gov/library/publications/the-world-factbook/rankorder/2001rank.html#pl

———, "Bulgaria," *World Factbook*, October 6, 2016. As of October 18, 2016:
https://www.cia.gov/library/publications/the-world-factbook/geos/bu.html

Central Statistics Bureau of Latvia, "ISG09. Population of Latvia by Citizenship at the Beginning of the Year," undated-a. As of March 2016:
http://data.csb.gov.lv/pxweb/en/Sociala/Sociala__ikgad__iedz__iedzskaits/IS0090.px/?rxid=a79839fe-11ba-4ecd-8cc3-4035692c5fc8

———, "NBG04. Activity Rate, Employment Rate and Unemployment Rate by Statistical Region," undated-b. As of March 2016:
http://data.csb.gov.lv/pxweb/en/Sociala/Sociala__ikgad__nodarb/NB0040.px/table/tableViewLayout1/?rxid=562c2205-ba57-4130-b63a-6991f49ab6fe

———, "Statistics Database," undated-c. As of April 2016:
http://www.csb.gov.lv/en/dati/statistics-database-30501.html

Charap, Samuel, "The Ghost of Hybrid War," *Survival*, Vol. 57, No. 6, December 2015, pp. 51–58. As of March 16, 2016:
http://scharap.fastmail.net/files/Ghost of Hybrid War.pdf

Chazan, David, "Russia 'Bought' Marine Le Pen's Support over Crimea," *The Telegraph*, April 4, 2015. As of March 15, 2016:
http://www.telegraph.co.uk/news/worldnews/europe/france/11515835/Russia-bought-Marine-Le-Pens-support-over-Crimea.html

Christides, Giorgos, "Could Europe Lose Greece to Russia?" BBC News, March 12, 2015. As of July 22, 2016:
http://www.bbc.com/news/world-europe-31837660

Cianetti, Licia, "The Governing Parties Survived Latvia's Election, but the Issue of the Country's Russian-Speaking Minority Remains Centre-Stage," London School of Economics blog, October 10, 2014. As of April 2016:
http://blogs.lse.ac.uk/europpblog/2014/10/08/the-governing-parties-survived-latvias-election-but-the-issue-of-the-countrys-russian-speaking-minority-remains-centre-stage/

Coalson, Robert, "New Greek Government Has Deep, Long-Standing Ties with Russian 'Fascist' Dugin," Radio Free Europe/Radio Liberty, January 22, 2015. As of July 22, 2016:
http://www.rferl.org/content/greek-syriza-deep-ties-russian-eurasianist-dugin/26818523.html

Cohen, Raphael S., and Gabriel M. Scheinmann, "Can Europe Fill the Void in U.S. Military Leadership?" *Orbis*, Vol. 58, No. 1, 2014, pp. 39–54.

"Concern Grows over Bulgarian Paramilitaries and 'Border Patrols,'" *Sofia Globe*, July 7, 2016. As of July 30, 2016:
http://sofiaglobe.com/2016/07/07/concern-grows-over-bulgarian-paramilitaries-and-border-patrols/

Conley, Heather, James Mina, Ruslan Stefanov, and Martin Vladmirov, *The Kremlin Playbook: Understanding Russian influence in Central and Eastern Europe*, Lanham, Md.: CSIS Europe Program and CSD Economics Program, Rowman and Littlefield, October 2016. As of February 2, 2018:
https://csis-prod.s3.amazonaws.com/s3fs-public/publication/1601017_Conley_KremlinPlaybook_Web.pdf

Connolly, Kate, "German Election: Merkel Wins Fourth Term but Far-Right AfD Surges to Third," *The Guardian*, September 24, 2017. As of January 9, 2018:
https://www.theguardian.com/world/2017/sep/24/angela-merkel-fourth-term-far-right-afd-third-german-election

"Conscious Uncoupling," *The Economist*, April 5, 2014. As of March 31, 2016:
http://www.economist.com/news/briefing/21600111-reducing-europes-dependence-russian-gas-possiblebut-it-will-take-time-money-and-sustained

Cousens, Elizabeth M., and Charles K. Cater, *Toward Peace in Bosnia: Implementing the Dayton Accords*, Boulder, Colo.: Lynne Rienner Publishers, 2001.

Covington, Stephen, *Putin's Choice for Russia*, Cambridge, Mass.: Belfer Center for Science and International Affairs, August 2015. As of October 26, 2016:
http://belfercenter.ksg.harvard.edu/files/Putins%20Choice%20web.pdf

"Customs Agency Audits Lukoil Neftochim," BTA, June 1, 2015. As of July 18, 2016:
http://www.bta.bg/en/c/DF/id/1091383

Darczewska, Jolanta, *The Anatomy of Russian Information Warfare: The Crimean Operation, A Case Study*, Warsaw, Poland: Centre for Eastern Studies, May 2014.

Dawisha, Karen, *Putin's Kleptocracy: Who Owns Russia?* New York: Simon and Schuster, 2014.

Day, Matthew, "'Extremely High' Number of Russian Spies in Czech Republic" *The Telegraph*, October 27, 2014. As of October 25, 2016:
http://www.telegraph.co.uk/news/worldnews/europe/czechrepublic/11190596/Extremely-high-number-of-Russian-spies-in-Czech-Republic.html

Deflin, Martin, "Podemos and the Iran-Venezuela Connection," *Deutsche Welle*, January 27, 2016. As of October 25, 2016:
http://www.dw.com/en/podemos-and-the-iran-venezuela-connection/a-19004706

DeGhett, Torie Rose, "Romania Is Starting to Freak Out About Russian Designs on Transnistria," Vice News, October 6, 2015. As of March 2016:
https://news.vice.com/article/romania-is-starting-to-freak-out-about-russian-designs-on-transnistria

Deliso, Christopher, "Analysis: LUKoil in the Balkans," UPI, June 26, 2002. As of July 18, 2016:
http://www.upi.com/Analysis-LUKoil-in-the-Balkans/57741025101874/

Demidoff, Maureen, "La Communauté Russe en France est «Éclectique [Russian Community in France Is Eclectic]," *Russie Info*, October 30, 2014. As of March 15, 2016:
http://www.russieinfo.com/
la-communaute-russe-en-france-est-eclectique-2014-10-30

Democratic Alliance of Hungarians in Romania, "Autonomy Does Not Mean Separation," June 16, 2014. As of July 2016:
http://www.dahr.ro/news/autonomy-does-not-mean-separation

Dempsey, Judy, "The Western Balkans Are Becoming Russia's New Playground," *Judy Dempsey's Strategic Europe*, Carnegie Europe, November 24, 2014. As of July 18, 2016:
http://carnegieeurope.eu/strategiceurope/?fa=57301

Destatis Statistiches Bundesamt, "Persons with a Migrant Background," undated. As of March 14, 2016:
https://www.destatis.de/EN/FactsFigures/SocietyState/Population/
MigrationIntegration/PersonsMigrationBackground/Tables/
MigrantStatusFormerCitizenhip.html

Die Linke, "Programme of the DIE LINKE Party," December 2011.

Dong, Phoebe, and Chris Rieser, "More Greeks Approve of Russia's Leadership Than EU's," Gallup, February 2, 2015. As of July 22, 2016:
http://www.gallup.com/poll/181460/greeks-approve-russia-leadership.aspx

Draper, Lucy, "Hackers Leak Messages 'Between Kremlin and France's Front National,'" *Newsweek*, April 3, 2015. As of March 15, 2016:
http://www.newsweek.com/
hackers-claim-leak-messages-between-kremlin-and-frances-front-national-319442

Dülffer, Meike, Carsten Luther, and Zacharias Zacharakis, "Caught in the Web of the Russian Ideologues," *Die Zeit*, February 7, 2015. As March 22, 2016:
http://pdf.zeit.de/politik/ausland/2015-02/russia-greece-connection-alexander-dugin-konstantin-malofeev-panos-kammeno.pdf

Dunlop, John B., "Aleksandr Dugin's Foundation of Geopolitics," *Demokratizatsiya*, Vol. 12, No. 1, January 31, 2004.

Economic Intelligence Unit, "Russian Spy Centre in Nis?" *The Economist*, November 21, 2014. As of July 18, 2016:
http://country.eiu.com/article.aspx?articleid=1152511099&Country=
Serbia&topic=Politics&subtopic=Forecast&subsubtopic=
International+relations&u=1&pid=394333423&oid=394333423&uid=1

Economides, Spyros, James Ker-Lindsay, and Dimitris Papadimitriou, "Kosovo: Four Futures," *Survival*, Vol. 52, No. 5, 2010, pp. 99–116.

Eke, Steven, "Russia President Targets Diaspora," BBC, October 24, 2006. As of March 16, 2015:
http://news.bbc.co.uk/2/hi/europe/6080928.stm

"Election 2015: Results," BBC News, undated. As of March 18, 2016:
http://www.bbc.com/news/election/2015/results

Estonia Internal Security Service, "Annual Review," 2013. As of October 26, 2016:
http://web.archive.org/web/20150404050206/https://www.kapo.ee/cms-data/_text/138/124/files/kapo-annual-review-2013-eng.pdf

———, "Annual Review," 2014. As of October 26, 2016:
https://www.kapo.ee/sites/default/files/public/content_page/Annual%20Review%202014.pdf

"Estonian Court Acquits Defendants in Bronze Soldier Protests," Sputnik International, May 1, 2009. As of April 2016:
https://sputniknews.com/world/20090105119376653/

"EU-Moscow Row over South Stream Gas Pipeline," BBC News, June 9, 2014. As of July 18, 2016:
http://www.bbc.com/news/world-europe-27767345

"Europe Far-Right Parties Meet in St. Petersburg, Russia," BBC News, March 22, 2015. As of March 15, 2016:
http://www.bbc.com/news/world-europe-32009360

European Commission, "Communication from the Commission to the European Parliament and the Council on the Short Term Resilience of the European Gas System Preparedness for a Possible Disruption of Supplies from the East During the Fall and Winter of 2014/2015," SWD(2014) 322 final, Brussels, October 16, 2014.

———, "EU Revenue and Expenditures," Excel file, 2015. As of July 2016:
http://ec.europa.eu/budget/revexp/revenue_and_expenditure_files/data/revenue_and_expenditure_en.xls

———, "Energy Security Strategy," March 31, 2016a. As of March 31, 2016:
https://ec.europa.eu/energy/node/192

———, "Imports and Secure Supplies: Diverse, Affordable, and Reliable Energy from Abroad," March 31, 2016b. As of March 31, 2016:
https://ec.europa.eu/energy/en/topics/imports-and-secure-supplies

European Commission Directorate-General for Economic and Financial Affairs, *European Economic Forecast, Winter 2016*, Institutional Paper 020, 2016. As of March 24, 2016:
http://ec.europa.eu/economy_finance/publications/eeip/pdf/ip020_en.pdf

European Council on Foreign Relations, "Public Opinion Poll: Bulgarian Foreign Policy, the Russia-Ukraine Conflict and National Security," March 26, 2015. As of July 19, 2016:
http://www.ecfr.eu/article/public_opinion_poll311520

European Union Agency for Fundamental Rights, *EU-MIDIS: European Union Minorities and Discrimination Survey*, Luxembourg: Publications Office of the European Union, December 2009. As of June 1, 2016:
http://fra.europa.eu/en/publication/2012/eu-midis-main-results-report

European Union External Action, "Normalisation of Relations between Belgrade and Pristina," undated. As of March 2016:
https://eeas.europa.eu/diplomatic-network/eu-facilitated-dialogue-belgrade-pristina-relations/349/dialogue-between-belgrade-and-pristina_en

Eurostat, "GDP Per Capita, Consumption Per Capita and Price Level Indices," *Eurostat Statistics Explained*, June 2016a. As of July 14, 20116:
http://ec.europa.eu/urope/statistics-explained/index.php/GDP_per_capita,_consumption_per_capita_and_price_level_indices

———, "Real GDP Growth Rate—Volume," July 11, 2016b. As of July 15, 2016:
http://ec.europa.eu/urope/tgm/table.do?tab=table&init=1&language=en&pcode=tec00115&plugin=1

Fairclough, Gordon, "EU and Russia Loom over Serbian Election," *Wall Street Journal*, May 4, 2012. As of July 19, 2016:
http://www.wsj.com/articles/SB10001424052702303990604577369812911390898

"Far-Right Trolls Active on Social Media Before German Election: Research," *Deutsche Welle*, February 21, 2018. As of March 2, 2018:
http://www.dw.com/en/far-right-trolls-active-on-social-media-before-german-election-research/a-42667318

Federal Agency for the Commonwealth of Independent States, Compatriots Living Abroad, and International Humanitarian Cooperation, "About *Rossotrudnichestvo*," webpage, undated. As of of March 23, 2016:
http://www.rs.gov.ru/en/about

"Ссора с россией обернулась для эстонии огромными потерями [Fight with Russia Brought Estonia Great Losses]," zagolovki.ru, November 17, 2007. As of April 2016:
http://zagolovki.ru/daytheme/urope/17Nov2007

Foster, Peter, and Matthew Holehouse, "Russia Accused of Clandestine Funding of European Parties as U.S. Conducts Major Review of Vladimir Putin's Strategy," *The Telegraph*, January 16, 2016. As of March 15, 2016:
http://www.telegraph.co.uk/news/worldnews/europe/russia/12103602/America-to-investigate-Russian-meddling-in-EU.html

Frear, Thomas, Łukasz Kulesa, and Ian Kearns, *Dangerous Brinkmanship: Close Military Encounters Between Russia and the West in 2014*, London: European Leadership Network, November 2014. As of April 2016:
http://www.europeanleadershipnetwork.org/medialibrary/2014/11/09/6375e3da/Dangerous%20Brinkmanship.pdf

Frederick, Bryan, Matthew Povlock, Stephen Watts, Miranda Priebe, and Edward Geist, *Assessing Russian Reactions to U.S. and NATO Posture Enhancements*, Santa Monica, Calif.: RAND Corporation, RR-1879-AF, 2017. As of February 24, 2018:
https://www.rand.org/pubs/research_reports/RR1879.html

"FSB Reveals It Had Agent in Estonia Intel for 20 Years," *Rossiyskaya Gazeta* via Interfax, December 14, 2014. As of April 2016:
http://rbth.com/news/2014/12/14/fsb_reveals_it_had_agent_in_estonia_intel_for_20_years_42235.html

Fukuyama, Francis, *The Origins of Political Order: From Prehuman Times to the French Revolution*, New York: Farrar, Straus and Giroux, 2014.

"Full Text: John Boehner Speaks at the Unveiling of Havel Bust," *Prague Post*, November 19, 2014.

Galeotti, Mark, "Organized Crime in the Baltic States," *Baltic Review*, March 24, 2015. As of April 2016:
http://baltic-review.com/organized-crime-in-the-baltic-states/

Gall, Carlotta, "How Kosovo Was Turned into Fertile Ground for ISIS" *New York Times*, May 21, 2016. As of July 2016:
http://www.nytimes.com/2016/05/22/world/europe/how-the-saudis-turned-kosovo-into-fertile-ground-for-isis.html?_r=1

Gauland, Alexander, "Thesenpapier Außenpolitik [Thesis on Foreign Policy]," *Alternative für Deutschland*, September 10, 2013. As of March 16, 2016:
https://www.alternativefuer.de/2013/09/11/thesenpapier-aussenpolitik/

Gertz, Bill, "Spy Ring Arrest Highlights Jump in Russian Spying Under Putin," *Washington Free Beacon*, January 28, 2015. As of March 15, 2016:
http://freebeacon.com/national-security/spy-ring-arrest-highlights-jump-in-russian-spying-under-putin/#

Giles, Keir, *Russia's 'New' Tools for Confronting the West Continuity and Innovation in Moscow's Exercise of Power*, London: Chatham House, March 2016. As of May 21, 2016:
https://www.chathamhouse.org/sites/files/chathamhouse/publications/research/2016-03-21-russias-new-tools-giles.pdf

"Global Opinion of Russia Mixed: Negative Views Widespread in Mideast and Europe," Pew Research Center, September 3, 2013. As of July 22, 2016:
http://www.pewglobal.org/2013/09/03/global-opinion-of-russia-mixed/

Goble, Paul, "Moscow Using Russian Organizations to Destabilize Latvia, Riga Officials Say," *The Interpreter*, March 10, 2015. As of April 6, 2016: http://www.interpretermag.com/moscow-using-russian-organizations-to-destabilize-latvia-riga-officials-say/

Goldman, Russell, "Russian Violations of Airspace Seen as Unwelcome Test by the West," *International New York Times*, October 6, 2015. As of March 31, 2016: http://www.nytimes.com/2015/10/07/world/europe/russian-violations-of-airspace-seen-as-unwelcome-test-by-the-west.html

Gorenburg, Dmitri, "Moscow Conference on International Security 2015, Part 2: Gerasimov on Military Threats Facing Russia," *Russian Military Reform* blog, May 4, 2015. As of January 2016: https://russiamil.wordpress.com/2015/05/04/moscow-conference-on-international-security-2015-part-2-gerasimov-on-military-threats-facing-russia/

Gray, Rosie, "Pro-Putin Think Tank Based in New York Shuts Down," BuzzFeed, June 30, 2015. As of March 23, 2016: http://www.buzzfeed.com/rosiegray/pro-putin-think-tank-based-in-new-york-shuts-down#.ld41qYzYm3

Greenberg, Andy, "The NSA Confirms It: Russia Hacked French Election Infrastructure," *Wired*, May 9, 2018. As of March 2, 2018: https://www.wired.com/2017/05/nsa-director-confirms-russia-hacked-french-election-infrastructure/

Greene, Sam, and Graeme Robertson, "Explaining Putin's Popularity: Rallying Around the Russian Flag," *Washington Post*, September 9, 2014. As of January 2016: https://www.washingtonpost.com/news/monkey-cage/wp/2014/09/09/explaining-putins-popularity-rallying-round-the-russian-flag/

Grigas, Agnia, "The New Generation of Baltic Russian Speakers," EurActiv.com, November 28, 2014. As of June 1, 2016: http://www.euractiv.com/sections/europes-east/new-generation-baltic-russian-speakers-310405

———, *Beyond Crimea: The New Russian Empire*, New Haven, Conn.: Yale University Press, 2016.

Groskop, Viv, "How the Ukraine Crisis Is Affecting Russians in Moscow-on-Thames," *The Guardian*, April 6, 2014. As of March 15, 2016: http://www.theguardian.com/world/2014/apr/06/among-the-russians-in-london

Gysi, Gregor, "Ukraine—Diplomacy Is the Only Way," *The Bullet*, Socialist Project E-Bulletin, No. 951, March 19, 2014. As March 22, 2016: http://www.socialistproject.ca/bullet/951.php#continue

Hanley-Giersch, Jennifer, "The Baltic States and the North Eastern European Criminal Hub," *ACAMS Today*, Association of Certified Anti-Money Laundering Specialists, September–November 2009, pp. 36–38. As of April 2016: http://www.globalriskaffairs.com/wp-content/uploads/2010/11/BalticStates_AT_2009-Kopie.pdf

Harding, Luke, "Gordievsky: Russia Has as Many Spies in Britain Now as the USSR Ever Did," *The Guardian*, March 11, 2013. As of March 15, 2016: http://www.theguardian.com/world/2013/mar/11/russian-spies-britain-oleg-gordievsky

Harrison, Colin, and Zuzana Princova, "A Quiet Gas Revolution in Central and Eastern Europe," *Energy Post*, October 29, 2015. As of July 15, 2016: http://www.energypost.eu/quiet-revolution-central-eastern-european-gas-market/

Hawley, Charles, "Merkel Re-Elected as Right Wing Enters Parliament," Spiegel Online International, September 24, 2017. As of March 2, 2018: http://www.spiegel.de/international/germany/german-election-merkel-wins-and-afd-wins-seats-in-parliament-a-1169587.html

Heickerö, Roland, *Emerging Cyber Threats and Russian Views on Information Warfare and Information Operations*, Stockholm, Sweden: FOI Swedish Defense Research Agency, 2010.

Helmus, Todd, Elizabeth Bodine-Baron, Andrew Radin, Madeline Magnuson, Joshua Mendelsohn, Bill Marcellino, Andriy Bega, and Zev Winkelman, *Russia Social Media Influence: Understanding Russian Propaganda in Eastern Europe*, Santa Monica, Calif.: RAND Corporation, RR-2237-OSD, 2018. As of June 8, 2018: https://www.rand.org/pubs/research_reports/RR2237.html

Hendrickson, Ryan C., "NATO's Visegrad Allies: The First Test in Kosovo," *Journal of Slavic Military Studies*, Vol. 13, No. 2, 2000, pp. 25–38.

Herszenhorn, David, "Facing Russian Threat, Ukraine Halts Plans for Deals with E.U.," *New York Times*, November 21, 2013. As of April 2016: http://www.nytimes.com/2013/11/22/world/europe/ukraine-refuses-to-free-ex-leader-raising-concerns-over-eu-talks.html?_r=0

Higgins, Andrew, "Far-Right Fever for a Europe Tied to Russia," *New York Times*, May 20, 2014a. As of March 18, 2016: http://www.nytimes.com/2014/05/21/world/europe/europes-far-right-looks-to-russia-as-a-guiding-force.html?_r=2

———, "Russian Money Suspected Behind Fracking Protests," *New York Times*, November 30, 2014b. As of July 18, 2016: http://www.nytimes.com/2014/12/01/world/russian-money-suspected-behind-fracking-protests.html?_r=0

Hill, Fiona, "Putin: The One-Man Show the West Doesn't Understand," *Bulletin of the Atomic Scientists*, Vol. 72, No. 3, 2016.

Hill, Fiona, and Clifford Gaddy, *Mr. Putin: Operative in the Kremlin*, Washington, D.C.: Brookings Institution, 2015.

Holbrooke, Richard, *To End a War*, New York: Modern Library, 1999.

"How the EU Lost Russia over Ukraine," Spiegel Online International, November 24, 2014. As of April 2016:
http://www.spiegel.de/international/europe/war-in-ukraine-a-result-of-misunderstandings-between-europe-and-russia-a-1004706-2.html

"How to Deal with Harmony," *The Economist*, October 5, 2014. As of April 2016 (subscription only):
http://www.economist.com/blogs/easternapproaches/2014/10/latvias-election?zid=307&ah=5e80419d1bc9821ebe173f4f0f060a07

Husain, Aasim M., Anna Ilyina, and Li Zeng, "Europe's Russian Connections," *VOX* (Centre for Economic Policy Research's Portal), August 29, 2014. As of July 15, 2016:
http://voxeu.org/article/europe-s-russian-connections

Ideon, Argo, "Expert: FSB Likely Fearing Capture of Vital Agent in Estonia," *Postimees*, December 30, 2014. As of April 2016:
http://news.postimees.ee/3039997/expert-fsb-likely-fearing-capture-of-vital-agent-in-estonia

IHS Janes, "External Affairs," *Jane's Sentinel Security Assessment*, August 14, 2014.

"In Moscow, Bulgarian Socialist Party Bemoans Sanctions Against Russia," *Sophia Globe*, March 18, 2015. As March 23, 2016:
http://sofiaglobe.com/2015/03/18/in-moscow-bulgarian-socialist-party-bemoans-sanctions-against-russia/

"In the Balkans, NATO Has Outmuscled Russia," *The Economist*, December 11, 2015. As of July 18, 2016:
http://www.economist.com/news/europe/21683967-montenegros-accession-fills-one-few-remaining-gaps-western-alliance

"In the Kremlin's Pocket: Who Backs Putin, and Why," *The Economist*, February 14, 2015. As of March 18, 2016:
http://www.economist.com/news/briefing/21643222-who-backs-putin-and-why-kremlins-pocket

Inspector General, U.S. Department of Defense, "Evaluation of the European Reassurance Initiative (ERI)," August 22, 2017. As of June 28, 2018:
https://media.defense.gov/2017/Sep/01/2001802392/-1/-1/1/DODIG-2017-111.PDF

Institute of Democracy and Cooperation, "The Institute of Democracy and Cooperation," undated. As of March 15, 2016:
http://www.idc-europe.org/en/The-Institute-of-Democracy-and-Cooperation

Internal Revenue Service, "Yearly Average Currency Exchange Rates," January 15, 2016. As of March 2016:
https://www.irs.gov/Individuals/International-Taxpayers/Yearly-Average-Currency-Exchange-Rates

International Center for Defense and Security, "Russia's Involvement in the Tallinn Disturbances," May 11, 2007. As of January 15, 2016:
http://www.icds.ee/publications/article/russias-involvement-in-the-tallinn-disturbances/

International Crisis Group, *Kosovo's Fragile Transition*, Europe Report, No. 196, September 25, 2008.

International Monetary Fund, *Central, Eastern, and Southeastern Europe: How to Get Back on the Fast Track*, May 2016. As of July 15, 2016:
https://www.imf.org/external/pubs/ft/reo/2016/eur/eng/pdf/rei0516.pdf

Ivanov, Igor, "The Missile-Defense Mistake: Undermining Strategic Stability and the ABM Treaty," *Foreign Affairs*, Vol. 79, No. 5, September–October 2000, pp. 15–20.

Ivanov, Vladimir, "Washington's Baltic Role: America Strengthens Eastern Flank of NATO," *The Independent* (in Russian), March 4, 2016. As of April 2016:
http://nvo.ng.ru/realty/2016-03-04/1_washington.html

"Jobbik MEP Béla Kovács, Accused of Spying for Russia Previously, to Lose Immunity," *Hungary Today*, October 13, 2015. As of March 18, 2016:
http://hungarytoday.hu/news/jobbik-mep-bela-kovacs-accused-spying-previously-lose-immunity-79900

"John Laughland," Hungarian Review, undated. As of March 23, 2016:
http://www.hungarianreview.com/author/john_laughland

Jones, Sam, "Estonia Ready to Deal with Little Green Men," *Financial Times*, May 13, 2015. As of April 2016:
http://www.ft.com/intl/cms/s/0/03c5ebde-f95a-11e4-ae65-00144feab7de.html

Jones, Sam, Kerin Hope, and Courtney Weaver, "Alarm Bells Ring over Syriza's Russian Links," *Financial Times*, January 28, 2015. As of March 21, 2016:
http://www.ft.com/intl/cms/s/0/a87747de-a713-11e4-b6bd-00144feab7de.html#axzz43YJBZZ6e

Judah, Tim, *Kosovo: War and Revenge*, New Haven, Conn.: Yale University Press, 2002.

———, *Kosovo: What Everyone Needs to Know*, Oxford, United Kingdom: Oxford University Press, 2008.

Jukic, Elvira M., "Russia Flexes Muscles on EU Bosnia Mission," *Balkan Insight*, November 17, 2014. As of July 2016:
http://www.balkaninsight.com/en/article/russia-flexes-muscles-on-eu-bosnia-mission

Kaiser, Robert, "Reassembling the Event: Estonia's 'Bronze Night,'" *Environment and Planning D: Society and Space*, Vol. 30, 2012, pp. 1051–1052.

Kalyvas, Stathis, *Logic of Violence in Civil Wars*, New York: Cambridge University Press, 2006.

Kaminski, Matthew, "The Town Where the Russian Dilemma Lives," *Wall Street Journal*, July 4, 2014. As of March 2016:
http://online.wsj.com/articles/
matthew-kaminski-the-town-where-the-russian-dilemma-lives-1404510023

Kanevskaya, Natalya, "How the Kremlin Wields Its Soft Power in France," Radio Free Europe/Radio Liberty, June 24, 2014. As of March 15, 2016:
http://www.rferl.org/content/russia-soft-power-france/25433946.html

Kantchev, Georgi, "With U.S. Gas, Europe Seeks Escape from Russia's Energy Grip," *Wall Street Journal*, February 25, 2016. As of March 31, 2016:
http://www.wsj.com/articles/
europes-escape-from-russian-energy-grip-u-s-gas-1456456892

KAPO—*See* Estonia Internal Security Service.

Kasekamp, Andres, *A History of the Baltics*, New York: Palgrave Macmillan, 2010.

Kassam, Ashifa, "Spanish Election: National Newcomers End Era of Two-Party Dominance," *The Guardian*, December 21, 2015. As of March 22, 2016:
http://www.theguardian.com/world/2015/dec/20/
peoples-party-wins-spanish-election-absolute-majority

Kaža, Juris, and Liis Kangsepp, "Baltic Countries Fear Impact of Russian Food Sanctions on Business," *Wall Street Journal*, August 7, 2014. As of June 2, 2016:
http://www.wsj.com/articles/
baltic-countries-fear-impact-of-russian-food-sanctions-on-business-1407437297

Kennan, George F., "Long Telegram" (Moscow to Washington), February 22, 1946, via National Security Archive. As of February 2016:
http://nsarchive.gwu.edu/coldwar/documents/episode-1/kennan.htm

Kennan, George F., in Giles D. Harlow and George C. Maerz, eds., *Measures Short of War: The George F. Kennan Lectures at the National War College, 1946-1947*, Washington, D.C.: National Defense University Press, 1991.

Kevin, Deirdre, Francesca Pellicanò, and Agnes Schneeberger, "Television News Channels in Europe," *European Audiovisual Observatory*, October 2013. As of March 23, 2016:
http://www.obs.coe.int/documents/205595/264629/
European+news+Market+2013+FINAL.pdf/
116afdf3-758b-4572-af0f-61297651ae80

Kirby, Paul, "Russia's Gas Fight with Ukraine," BBC News, October 31, 2014. As of April 2016:
http://www.bbc.com/news/world-europe-29521564

Kirch, Aksel, Marika Kirch, and Tarmo Tuisk, "Russians in the Baltic States: To Be or Not to Be?" *Journal of Baltic Studies*, Vol. 24, No. 2, Summer 1993, pp. 173–188.

Kirk, Ashely, "What Are the Biggest Defence Budgets in the World?" *The Telegraph*, October 27, 2015. As of March 10, 2016:
http://www.telegraph.co.uk/news/uknews/defence/11936179/What-are-the-biggest-defence-budgets-in-the-world.html

———, "Iraq and Syria: How Many Foreign Fighters are Fighting for ISIL?" *The Telegraph*, March 24, 2016. As of March 30, 2016:
http://www.telegraph.co.uk/news/worldnews/islamic-state/11770816/Iraq-and-Syria-How-many-foreign-fighters-are-fighting-for-Isil.html

Kirschbaum, Erik, "Putin's Apologist? Germany's Schroeder Says They're Just Friends," Reuters, March 27, 2014. As of March 11, 2016:
http://www.reuters.com/article/ukraine-russia-schroeder-idUSL5N0MN3ZI20140327

Kivirähk, Juhan, *Integrating Estonia's Russian-Speaking Population: Findings of National Defense Opinion Surveys*, International Centre for Defence and Security, December 2014.

———, *Public Opinion and National Defence*, Estonian Ministry of Defence, April 2015. As of April 2016:
http://www.kaitseministeerium.ee/sites/default/files/elfinder/article_files/public_opinion_and_national_defence_2015_march_0.pdf

Klapsis, Antonis, *An Unholy Alliance: The European Far Right and Putin's Russia*, Brussels: Wilfried Marten Centre for European Studies, 2015. As of March 21, 2016:
http://www.martenscentre.eu/sites/default/files/publication-files/far-right-political-parties-in-europe-and-putins-russia.pdf

Knezevic, Gordana, "Wanting the Best of Both Worlds: How Serbs View Russia and EU," Radio Free Europe/Radio Liberty, May 13, 2016. As of July 19, 2016:
https://www.ceas-serbia.org/sr/preuzeto/wanting-the-best-of-both-worlds-how-serbs-view-russia-and-eu

Knight, Ben, "After the Gold Rush: AfD Loses State Subsidies," *Deutsche Welle*, December 18, 2015. As of March 16, 2016:
http://www.dw.com/en/after-the-gold-rush-afd-loses-state-subsidies/a-18928520

Knoema, "Leningrad Region—Gross Regional Product per Capita," undated-a. As of April 2016:
http://knoema.com/atlas/Russian-Federation/Leningrad-Region/GRP-per-capita

———, "Pskov Region—Gross Regional Product per Capita," undated-b. As of April 2016:
http://knoema.com/atlas/Russian-Federation/Pskov-Region/topics/Gross-regional-product/Gross-regional-product/GRP-per-capita

"Kobzon Thanked Siderov for His Position on Behalf of Russia and Wished ATAKA Success at the Elections," ATAKA, May 22, 2014.

Kofman, Michael, "Russian Hybrid Warfare and Other Dark Arts," *War on the Rocks*, March 11, 2016. As of April 2016:
http://warontherocks.com/2016/03/russian-hybrid-warfare-and-other-dark-arts/

Kofman, Michael, Katya Migacheva, Brian Nichiporuk, Andrew Radin, Olesya Tkacheva, and Jenny Oberholtzer, *Lessons from Russia' Operations in Crimea and Eastern Ukraine*, Santa Monica, Calif.: RAND Corporation, RR-1498-A, 2017. As of June 26, 2018:
https://www.rand.org/pubs/research_reports/RR1498.html

Kotkin, Stephen, "Russia's Perpetual Geopolitics: Putin Returns to Historical Patterns," *Foreign Affairs*, May/June 2016. As of July 2016:
https://www.foreignaffairs.com/articles/ukraine/2016-04-18/russias-perpetual-geopolitics

Kovacevic, Danijel, "Bosnian Serb Leader Postpones Controversial Referendum," *Balkan Insights*, February 9, 2016. As of July 2016:
http://www.balkaninsight.com/en/article/bosnian-serb-leader-puts-controversial-referendum-on-hold--02-09-2016

Kropaite, Aivile, "Baltic States Count Cost of Ending Soviet Electricity Link," *EU Observer*, December 15, 2015. As of June 2, 2016:
https://euobserver.com/economic/131524

Kudors, Andis, "Reinventing View to the Russian Media and Compatriot Policy in the Baltic States," in Artis Pabriks and Andis Kudors, eds., *The War in Ukraine: Lessons for Europe*, Riga, Latvia: Centre for Eastern European Policy Studies, University of Latvia Press, 2015.

Kukk, Kadri, "Brief History of 'Night Watch' in Estonia," Café Babel, May 7, 2007. As of April 6, 2016:
http://www.cafebabel.co.uk/tallinn/article/brief-history-of-nightwatch-in-estonia.html

Laitin, David, *Identity in Formation*, Ithaca, N.Y.: Cornell University Press, 1998.

Larrabee, F. Stephen, Stephanie Pezard, Andrew Radin, Nathan A. Chandler, Keith W. Crane, and Thomas S. Szayna, *Russia and the West After the Ukraine Crisis: European Vulnerabilities to Russian Pressures*, Santa Monica, Calif.: RAND Corporation, RR-1305-A, 2017. As of June 25, 2018:
https://www.rand.org/pubs/research_reports/RR1305.html

Laruelle, Marlene, *Russian Eurasianism: Ideology of Empire*, Washington, D.C.: Woodrow Wilson Center Press, 2012.

———, *The 'Russian World': Russia's Soft Power and Geopolitical Imagination*, Washington, D.C.: Center on Global Interests, May 2015.

Latvia Security Police, "Annual Report," 2013. As of October 26, 2016:
http://dp.gov.lv/en/?rt=documents&ac=download&id=6

Latvijas Gāze, homepage, undated. As of February 1, 2018:
http://www.lg.lv/?id=194&lang=eng

Lavrov, Sergey, "Russia's Foreign Policy: Historical Background," *Russia in Global Affairs*, via Russian Federation Ministry of Foreign Affairs website, March 3, 2016. As of March 26, 2016:
http://www.mid.ru/en/foreign_policy/news/-/asset_publisher/cKNonkJE02Bw/content/id/2124391/pop_up?_101_INSTANCE_cKNonkJE02Bw_viewMode=print&_101_INSTANCE_cKNonkJE02Bw_qrIndex=0

Lensky, Maxim, and Nikolai Flichenko, "Tallinn: A Year Without the Bronze Soldier," *Kommersant*, April 25, 2008. As of April 2016:
http://www.kommersant.com/p886482/r_1/foreign_relations/

Leonhard, Robert R., and Stephen P. Phillips, *'Little Green Men': A Primer on Modern Russian Unconventional Warfare*, Fort Bragg, N.C.: U.S. Army Special Operations Command, Assessing Revolutionary and Insurgent Strategies Project, undated. As of October 26, 2016:
http://www.jhuapl.edu/Content/documents/ARIS_LittleGreenMen.pdf

Leviev-Sawyer, Clive, "Pan-Orthodox Council: Bulgarian, Russian Orthodox Church Positions 'Overlap,'" *Sofia Globe*, June 10, 2016. As of June 2016:
http://sofiaglobe.com/2016/06/10/pan-orthodox-council-bulgarian-russian-orthodox-church-positions-overlap/

Lukyanov, Fyodor, "The Lost Twenty-Five Years," *Russia in Global Affairs*, February 28, 2016. As of March 2016:
http://eng.globalaffairs.ru/redcol/The-Lost-Twenty-Five-Years-18012

Lyman, Rick, and Helene Bienvenufeb, "Hungary Keeps Visit by Putin Low-Key as It Seeks to Repair Relations with West," *New York Times*, February 17, 2015. As of October 25, 2016:
http://www.nytimes.com/2015/02/18/world/hungary-keeps-visit-by-putin-low-key-as-it-seeks-to-repair-relations-with-west.html

Lyon, James, "Is War About to Break Out in the Balkans?" *Foreign Policy*, October 26, 2015. As of July 2016:
http://foreignpolicy.com/2015/10/26/war-break-out-balkans-bosnia-republika-srpska-dayton/

MacFarquhar, Neil, "Russia Revisits Its History to Nail Down Its Future," *New York Times*, May 11, 2014. As of March 23, 2016:
http://www.nytimes.com/2014/05/12/world/europe/russia-revisits-its-history-to-nail-down-its-future.html?_r=0

Maigre, Merle, "Nothing New in Hybrid Warfare," German Marshall Fund of the United States, policy brief, February 12, 2015.

Mankoff, Jeffrey, *Russian Foreign Policy: The Return of Great Power Politics*, Lanham, Md.: Rowman & Littlefield, 2012.

Marinas, Radu Sorin, "U.S. Raises Fresh Concerns Over Gazprom-Led Nord Stream 2," Reuters, February 18, 2016. As of March 31, 2016: http://www.reuters.com/article/us-energy-nordstream-usa-idUSKCN0VR25J

Martyn-Hemphill, Richard, "Estonia's New Premier Comes from Party with Links to Russia," *New York Times*, November 20, 2016. As of March 16, 2018: https://www.nytimes.com/2016/11/21/world/europe/estonia-juri-ratas-center-party.html

Mauricas, Žygimantas, "The Effect of Russian Economic Sanctions on Baltic States," Nordea Markets, undated. As of April 2016: https://nexus.nordea.com/research/attachment/17231

———, "Lithuania: Economic Dependence of Russia," Nordea Markets, March 20, 2014. As of April 2016: https://nexus.nordea.com/research/attachment/7286

McGuinness, Damien, "Russia Steps into Berlin 'Rape' Storm Claiming German Cover-Up," BBC News, January 27, 2016a. As of March 15, 2016: http://www.bbc.com/news/blogs-eu-35413134

———, "Germany Jolted by AfD Right-Wing Poll Success," BBC News, March 14, 2016b. As of March 15, 2016: http://www.bbc.com/news/world-europe-35806047

"Memorandum of Understanding (MoU) Between Ministry of Economic Affairs and Communications of the Republic of Estonia and Ministry of Defence of the Republic of Latvia and Ministry of National Defence of the Republic of Lithuania," November 4, 2015.

Mezzofiore, Gianluca, "Marine Le Pen's Front National Borrows €9m from Russian Lender," *International Business Times*, November 24, 2015. As of March 15, 2016: http://www.ibtimes.co.uk/marine-le-pens-front-national-borrows-9m-russian-lender-1476295

Michel, Casey, "Putin's Magnificent Messaging Machine," *Politico*, August 26, 2015. As of March 23, 2016: http://www.politico.eu/article/putin-messaging-machine-propaganda-russia-today-media-war/

"Migrant Crisis: Migration to Europe Explained in Seven Charts," BBC News, March 4, 2016. As of March 30, 2016: http://www.bbc.com/news/world-europe-34131911

Migranyan, Andranik, "Testimony on Russian-American Relations on the Question of Chechnya Before the U.S. Congress," House Committee on Foreign Affairs, Subcommittee on Europe, Eurasia, and Emerging Threats, April 26, 2013. As of March 23, 2016:
http://docs.house.gov/meetings/FA/FA14/20130426/100777/HHRG-113-FA14-Wstate-MigranyanA-20130426.pdf

Mihajlovic, Branka, "Russian Seeking Serbian Media Outlet?" Radio Slobodna Europa, February 14, 2016. As of July 14, 2016:
http://www.slobodnaevropa.org/a/russian-seeking-serbia-media-outlet/27551794.html

Milić, Jelena, "Putin's Orchestra," *The New Century*, No. 7, May 2014. As of July 18, 2016:
https://www.ceas-serbia.org/images/Jelena_Milic_Foreword_Putins_Orchestra.pdf

Milne, Richard, "Party with Ties to Putin Pushes Ahead in Estonian Polls," *Financial Times*, February 27, 2015. As of April 2016 (subscription only):
http://www.ft.com/intl/cms/s/0/1decfbac-be8a-11e4-a341-00144feab7de.html#axzz3hEUPUbiQ

Ministry of Tourism, Bulgaria, "In the Russian Duma: Bulgaria Is the Best Place to Accept Russian Tourists in 2016," February 25, 2016. As of July 18, 2016:
http://www.tourism.government.bg/en/kategorii/novini/russian-duma-bulgaria-best-place-accept-russian-tourists-2016

"Mystery Website Producer Has Ties to Harmony, LTV Reports" Latvian Public Broadcasting English-Language Service, January 9, 2017. As of March 16, 2018:
http://eng.lsm.lv/article/society/defense/mystery-websiteproducer-has-ties-to-harmony-ltv-reports.a218258/

Nasralla, Shadia, "Austrian Far-Right Party Gets Electoral Boost from Migrant Crisis," Reuters, September 27, 2015. As March 18, 2016:
http://www.reuters.com/article/us-austria-election-idUSKCN0RR0ZQ20150927

NATO—*See* North Atlantic Treaty Organization.

NATO Cooperative Cyber Defence Centre of Excellence, homepage, undated. As of April 2016:
https://ccdcoe.org

NATO Energy Security Center of Excellence, homepage, undated. As of February 1, 2018:
https://enseccoe.org/en/

North Atlantic Treaty Organization, "Study on NATO Enlargement," September 3, 1995. As of July 2016:
http://www.nato.int/cps/en/natohq/official_texts_24733.htm

———, "NATO's Assessment of a Crisis and Development of Response Strategies." June 16, 2011. As of March 24, 2016:
http://www.nato.int/cps/en/natohq/official_texts_75565.htm?

———, "Relations with Montenegro," May 26, 2016. As of July 2016:
http://www.nato.int/cps/en/natohq/topics_49736.htm

Observatory of Economic Complexity, "Russia," webpage, undated. As of February 1, 2018:
https://atlas.media.mit.edu/en/profile/country/rus/

Office of the Director of National Intelligence, *Assessing Russian Activities and Intentions in Recent US Elections*, Washington, D.C.: Intelligence Community Assessment, ICA 2017-01D, January 6, 2017. As of February 1, 2018:
https://www.dni.gov/files/documents/ICA_2017_01.pdf

Oja, Kaspar, "No Milk for the Bear: The Impact on the Baltic States of Russia's Counter-Sanctions," *Baltic Journal of Economics*, Vol. 15, No. 1, 2015, pp. 38–49.

Oldberg, Ingmar, *Kaliningrad's Difficult Plight Between Moscow and Europe*, Stockholm: Swedish Institute of International Affairs, No. 2, 2015. As of June 1, 2016:
https://www.ui.se/globalassets/butiken/ui-paper/2015/kaliningrads-difficult-plight-between-moscow-and-europe---io.pdf

Oliker, Olga, Christopher S. Chivvis, Keith Crane, Olesya Tkacheva, and Scott Boston, *Russian Foreign Policy in Historical and Current Context*, Santa Monica, Calif.: RAND Corporation, PE-144-A, 2015. As of March 2016:
http://www.rand.org/pubs/perspectives/PE144.html

Oliker, Olga, Keith Crane, Lowell H. Schwartz, and Catherine Yusupov, *Russian Foreign Policy: Sources and Implications*, Santa Monica, Calif.: RAND Corporation, MG-768-AF, 2009. As of March 16, 2016:
http://www.rand.org/pubs/monographs/MG768.html

Oliphant, Roland, "Mapped: Just How Many Incursions into NATO Airspace Has Russian Military Made?" *The Telegraph*, May 15, 2015. As of March 31, 2016:
http://www.telegraph.co.uk/news/worldnews/europe/russia/11609783/Mapped-Just-how-many-incursions-into-Nato-airspace-has-Russian-military-made.html

———, "Russia 'Simulated a Nuclear Strike' Against Sweden, NATO Admits," *The Telegraph*, February 4, 2016. As of March 31, 2016:
http://www.telegraph.co.uk/news/worldnews/europe/russia/12139943/Russia-simulated-a-nuclear-strike-against-Sweden-Nato-admits.html

OMV, "OMV and Gazprom Celebrate 40 Years of Natural Gas Import from Russia into Austria," April 17, 2008. As of March 31, 2016:
http://www.omv.com/portal/generic-list/display?lang=en&contentId=120835927935266

"Orbán: Keleti szél fúj [Orbán: Eastern Wind Is Blowing]," *Index* [Budapest], November 5, 2011.

Orenstein, Mitchell A., "Putin's Western Allies," *Foreign Affairs*, March 26, 2014. As of March 21, 2016:
https://www.foreignaffairs.com/articles/russia-fsu/2014-03-25/putins-western-allies

Orenstein, Mitchell A., Péter Krekó, and Attila Juhász, "The Hungarian Putin?" *Foreign Affairs*, February 8, 2015. As of July 2016:
https://www.foreignaffairs.com/articles/hungary/2015-02-08/hungarian-putin

Osborn, Andrew, "Paris Attacks, Hollande Visit May Spur Kremlin Push to End Isolation," Reuters, November 18, 2015. As of March 30, 2016:
http://www.reuters.com/article/
us-russia-putin-rehabilitation-idUSKCN0T72JA20151118

Osburg, Jan, *Unconventional Options for the Defense of the Baltic States: The Swiss Approach*, Santa Monica, Calif.: RAND Corporation, PE-179-RC, 2016. As of October 2016:
http://www.rand.org/pubs/perspectives/PE179.html

Ostroukh, Andrey, "Russian Economy Seen Growing From 2017 Onwards: World Bank," Reuters, May 23, 2017. As of January 9, 2018:
https://www.reuters.com/article/us-russia-worldbank/russian-economy-seen-growing-from-2017-onwards-world-bank-idUSKBN18J1DK?il=0

Ovozi, Qishloq, "The European Union, The Southern Corridor, and Turkmen Gas," Radio Free Europe/Radio Liberty, April 23, 2015. As of March 31, 2016:
http://www.rferl.org/content/
energy-pipelines-turkmenistan-former-soviet-europe/26974648.html

Ozimko, Kirill, "Is Russia Losing Its Little Brother? Information War Drags Serbia Closer to EU," *Fort Russ News*, December 20, 2015. As of June 2016:
http://www.fort-russ.com/2015/12/is-russia-losing-its-little-brother.html

Parkinson, Joe, "Bulgaria's Western Allies Worry About Eastern Tilt," *Wall Street Journal*, May 30, 2014. As of June 2016:
http://www.wsj.com/articles/
bulgarias-western-allies-worry-about-eastward-tilt-1401477681

Patrikarakos, David, "The Greeks Are Not 'Western:' Greece and Russia Breathe New Life into Their Ancient Eastern Alliance," *Politico*, April 22, 2015. As of March 21, 2016:
http://www.politico.eu/article/the-greeks-are-not-western/

Patterson, Tony, "Putin's Far-Right Ambition: Think-Tank Reveals How Russian President Is Wooing—and Funding—Populist Parties Across Europe to Gain Influence in the EU," *The Independent*, November 25, 2014. As March 16, 2016:
http://www.independent.co.uk/news/world/europe/
putin-s-far-right-ambition-think-tank-reveals-how-russian-president-is-wooing-and-funding-populist-9883052.html

Pavlovsky, Gleb, "Putin's World Outlook," *New Left Review*, No. 88, July–August 2014, pp. 55–66.

Perritt, Henry H., *The Road to Independence for Kosovo: A Chronicle of the Ahtisaari Plan*, Cambridge, United Kingdom: Cambridge University Press, 2009.

Person, Robert, "6 Reasons Not to Worry About Russia Invading the Baltics," *Washington Post*, November 12, 2015. As of June 1, 2016:
https://www.washingtonpost.com/news/monkey-cage/wp/2015/11/12/6-reasons-not-to-worry-about-russia-invading-the-baltics/

Pelnens, Gatis, ed., *The 'Humanitarian Dimension' of Russian Foreign Policy Toward Georgia, Moldova, Ukraine, and the Baltic States*, Riga, Latvia: Centre for East European Policy Studies, International Centre for Defence Studies, Centre for Geopolitical Studies, School for Policy Analysis at the National University of Kyiv-Mohyla Academy, Foreign Policy Association of Moldova, International Centre for Geopolitical Studies, 2009. As of July 5, 2016:
https://www.academia.edu/1376506/The_Humanitarian_Dimension_of_Russian_Foreign_Policy_Toward._Georgia_Moldova_Ukraine_and_the_Baltic_States

Pezard, Stephanie, Andrew Radin, Thomas Szayna, and F. Stephen Larrabee, *European Relations with Russia: Threat Perceptions, Responses and Strategies in the Wake of the Ukraine Crisis*, Santa Monica, Calif.: RAND Corporation, RR-1579-A, 2017. As of March 1, 2018:
https://www.rand.org/pubs/research_reports/RR1579.html

Pismennaya, Evgenia, and Anna Andrianova, "Russia's Economy Is Tanking, So Why Is Putin Smiling?" Bloomberg Business, February 29, 2016. As of March 29, 2016:
http://www.bloomberg.com/news/articles/2016-03-01/as-russia-s-economy-contracts-putin-s-preferred-indicator-is-up

Phillipson, Alice, "Berlusconi Says Vladimir Putin Wants Him to Become Russia's Economy Minister," *The Telegraph*, July 23, 2015. As of March 11, 2016:
http://www.telegraph.co.uk/news/worldnews/europe/italy/11758227/Berlusconi-says-Vladimir-Putin-wants-him-to-become-Russias-economy-minister.html

Poleshchuk, Vadim, *Russian-Speaking Population of Estonia in 2014: Monitoring Report*, Tallinn, Estonia: Legal Information Centre for Human Rights, 2014.

Pomerantsev, Peter, *Nothing Is True and Everything Is Possible: The Surreal Heart of the New Russia*, New York: Public Affairs, 2015.

Pukhov, Ruslan, "Nothing 'Hybrid' About Russia's War in Ukraine," *Moscow Times*, May 27, 2015. As of April 2016:
http://www.themoscowtimes.com/opinion/article/nothing-hybrid-about-russia-s-war-in-ukraine/522471.html

Putin, Vladimir, "Annual Address to the Federal Assembly of the Russian Federation," Moscow, April 25, 2005.

———, "Interview to *Politika* Newspaper," October 15, 2014. As of June 2016: http://en.kremlin.ru/events/president/news/46806

———, "Speech and the Following Discussion at the Munich Conference on Security Policy," Munich, Germany, February 12, 2007. As of October 26, 2016: http://web.archive.org/web/20160321232901/http://archive.kremlin.ru/eng/speeches/2007/02/10/0138_type82912type82914type82917type84779_118123

"Putin Promises Serbian Leader Russia Will Back Claim to Kosovo," Radio Free Europe/Radio Liberty, March 11, 2016. As of July 2016: http://www.rferl.org/content/putin-promises-serbian-leader-russia-will-back-claim-kosovo/27604028.html

"Putin: Romania 'in Crosshairs' After Opening NATO Missile Defense Base," RT, May 27, 2016. As of June 2016: https://www.rt.com/news/344642-putin-visit-greece-tsipras/

"Putin's Prepared Remarks at 43rd Munich Conference on Security Policy," transcript via *Washington Post*, February 12, 2007. As of March 23, 2016: http://www.washingtonpost.com/wp-dyn/content/article/2007/02/12/AR2007021200555.html

Radin, Andrew, "Towards the Rule of Law in Kosovo: Why EULEX Should Go," *Nationalities Papers*, Vol. 42, No. 2, March 2014, pp. 181–194.

———, *Hybrid Warfare in the Baltics: Threats and Potential Responses*, Santa Monica, Calif.: RAND Corporation, RR-1577-AF, 2017a. As of January 16, 2018: https://www.rand.org/pubs/research_reports/RR1577.html

———, "How NATO Could Accidentally Trigger a War with Russia," *The National Interest*, November 11, 2017b. As of January 16, 2018: http://nationalinterest.org/feature/how-nato-could-accidentally-trigger-war-russia-23156

Radin, Andrew, and Clinton Bruce Reach, *Russian Views of the International Order*, Santa Monica, Calif.: RAND Corporation, RR-1826-OSD, 2017. As of September 25, 2017: https://www.rand.org/pubs/research_reports/RR1826.html

Ragozin, Leonid, "Putin's Hand Grows Stronger as Right-Wing Parties Advance in Europe," Bloomberg News, March 15, 2016. As of March 23, 2016: http://www.bloomberg.com/news/articles/2016-03-15/putin-s-hand-grows-stronger-as-right-wing-parties-advance-in-europe

Rapoz, Kenneth, "How Lithuania Is Kicking Russia to the Curb," *Forbes*, October 18, 2015. As of June 2, 2016: http://www.forbes.com/sites/kenrapoza/2015/10/18/how-lithuania-is-kicking-russia-to-the-curb/4/#178da5ee251d

"Ratas: Center Party Not Planning to Give Up Protocol with United Russia," ERR.ee, October 9, 2017. As of March 15, 2018:
https://news.err.ee/635273/
ratas-center-party-not-planning-to-give-up-protocol-with-united-russia

Rettman, Andrew, "Western Balkans: EU Blindspot on Russian Propaganda," *EU Observer*, December 10, 2015. As of July 15, 2016:
https://euobserver.com/foreign/131472

———, "EU Warned of Russian 'Peril' in Western Balkans," *EU Observer*, July 12, 2016. As of July 14, 2016:
https://euobserver.com/foreign/134313

Reynolds, Paul, "New Russian World Order, The Five Principles," BBC News, September 1, 2008. As of March 2016:
http://news.bbc.co.uk/2/hi/europe/7591610.stm

Richter, Jan, "Miloš Zeman—Political Veteran Seeking to Crown His Career," Radio Praha, December 19, 2012. As of March 18, 2016:
http://www.radio.cz/en/section/curraffrs/
milos-zeman-political-veteran-seeking-to-crown-his-career?set_default_version=1

Rivera, Sharon Werning, James Byan, Brisa Camacho-Lovell, Carlos Fineman, Nora Klemmer, and Emma Raynor, *The Russian Elite 2016: Perspectives on Foreign and Domestic Policy*, Clinton, N.Y.: Arthur Levitt Public Affairs Center, Hamilton College, May 11, 2016.

Robinson, Linda, Todd C. Helmus, Raphael S. Cohen, Alireza Nader, Andrew Radin, Madeline Magnuson, and Katya Migacheva, *Modern Political Warfare: Current Practices and Possible Responses*, Santa Monica, Calif.: RAND Corporation, RR-1772-A, 2018. As of June 25, 2018:
https://www.rand.org/pubs/research_reports/RR1772.html

Robinson, Linda, Paul D. Miller, John Gordon IV, Jeffrey Decker, Michael Schwille, and Raphael S. Cohen, *Improving Strategic Competence: Lessons from 13 Years of War*, Santa Monica, Calif.: RAND Corporation, RR-816-A, 2014. As of July 20, 2016:
http://www.rand.org/pubs/research_reports/RR816.html

Ron Paul Institute for Peace and Prosperity, "About the Institute," undated. As of March 23, 2016:
http://ronpaulinstitute.org/about-us/

Roth, Andrew, "Former Russian Rebels Trade War in Ukraine for Posh Life in Moscow," *Washington Post*, September 16, 2015. As of March 2016:
https://www.washingtonpost.com/world/
former-russian-rebels-trade-war-in-ukraine-for-posh-life-in-moscow/2015/09/13/
6b71f862-3b8c-11e5-b34f-4e0a1e3a3bf9_story.html

"Russia and Estonia 'Exchange Spies' After Kohver Row," BBC, September 26, 2015. As of March 2016:
http://www.bbc.com/news/world-europe-34369853

"Russia Gives Blankets to Kosovo Serbs Instead of Citizenship," *Pravda*, December 8, 2011.

"Russia: Old Foe or New Ally?" Al Jazeera, December 12, 2015. As of March 23, 2016:
http://www.aljazeera.com/programmes/headtohead/2015/12/russia-foe-ally-151201112017212.html

"Russia's S-300 Missile Systems 'Too Costly' for Serbia—PM," Tass Russian News Agency, January 11, 2016. As of July 2016:
http://tass.ru/en/defense/848821

Russian Federation, National Security Strategy, Moscow, 2016.As of August 8, 2018:
http://www.ieee.es/Galerias/fichero/OtrasPublicaciones/Internacional/2016/Russian-National-Security-Strategy-31Dec2015.pdf

Russian Federation, Interdepartmental Foreign Policy Commission of the Security Council, Russian Foreign Policy Concept, April 1993.

Russian Federation, Ministry of Foreign Affairs, Concept of the Foreign Policy of the Russian Federation, February 12, 2013. As of April 25, 2016:
http://archive.mid.ru//brp_4.nsf/0/76389FEC168189ED44257B2E0039B16D

Russian Federation, Ministry of Foreign Affairs, Foreign Policy Concept of the Russian Federation, November 30, 2016. As of February 5, 2018:
http://www.mid.ru/en/foreign_policy/official_documents/-/asset_publisher/CptICkB6BZ29/content/id/2542248

Русская Школа Эстонии [Russian School of Estonia], homepage, undated. As of March 2016:
http://www.venekool.eu

Russkiy Mir, homepage, undated. As of April 2016:
http://russkiymir.ru/en/grants/index.php

Šajkaš, Marija, and Milka Tadić Mijović, "Caught Between the East and West: The "Media War" Intensifies in Serbia and Montenegro," Washington, D.C.: Center for International Media Assistance, National Endowment for Democracy, March 10, 2016. As of July 14, 2016:
http://www.cima.ned.org/blog/serbia-and-montenegro/

Sanger, David E., and Eric Schmidt, "Russian Ships Near Data Cables Are Too Close for U.S. Comfort," *International New York Times*, October 25, 2015. As of March 31, 2016:
http://www.nytimes.com/2015/10/26/world/europe/russian-presence-near-undersea-cables-concerns-us.html?_r=0

Scally, Derek, "Election of Die Linke State Premier Causes Stir in Germany," *Irish Times*, December 5, 2014. As of March 22, 2016:
http://www.irishtimes.com/news/world/europe/
election-of-die-linke-state-premier-causes-stir-in-germany-1.2027892

Schaefer, Olya, "Estonian Transit Shows Growth but Fears Russia," *Baltic Times*, March 9, 2011. As of April 2016:
http://www.baltictimes.com/news/articles/28207/

Scherer, Steve, "Italy's Berlusconi Says Crimea Split from Ukraine Was Democratic," Reuters, September 27, 2015. As of March 11, 2016:
http://www.reuters.com/article/
us-ukraine-crisis-berlusconi-idUSKCN0RR0MW20150927

Schmid, Fidelius, and Andreas Ulrich, "Betrayer and Betrayed: New Documents Reveal Truth on NATO's 'Most Damaging' Spy," Spiegel Online International, May 6, 2010. As of April 24, 2016:
http://www.spiegel.de/international/europe/betrayer-and-betrayed-new-documents-reveal-truth-on-nato-s-most-damaging-spy-a-693315.html

Schreck, Carl, "Russian Expats Wrestle with Dual-Citizenship Dilemma," Radio Free Europe/Radio Liberty, March 14, 2014. As of March 15, 2016:
http://www.rferl.org/content/russia-expatriates-dual-citizenship-law/25432010.html

Schwirtz, Michael, "German Election Mystery: Why No Russian Meddling?" *New York Times*, September 21, 2017. As of March 2, 2018:
https://www.nytimes.com/2017/09/21/world/europe/german-election-russia.html

Sengupta, Somini, "Russia Vetoes U.N. Resolution Calling Srebrenica Massacre 'Crime of Genocide,'" *New York Times*, July 8, 2015. As of July 2016:
https://www.nytimes.com/2015/07/09/world/europe/
russia-vetoes-un-resolution-calling-srebrenica-massacre-crime-of-genocide.html

"Serbian Delegates Say They Recognize Crimea Part of Russia," Tass Russian News Agency, October 27, 2015. A of July 19, 2016:
http://tass.ru/en/world/832067

Sevastopulo, Demetri, "Russian Navy Presents the US with a Fresh Challenge," *Financial Times*, November 2, 2015. As of March 30, 2016:
http://www.ft.com/intl/cms/s/0/
47314ece-80a8-11e5-8095-ed1a37d1e096.html#axzz44QIMBcVH

Sharkov, Damien, "Far-Right MEP Accused of Acting as Russian Spy," *Newsweek*, September 26, 2014. As of March 18, 2016:
http://www.newsweek.com/far-right-mep-accused-acting-russian-spy-273444

Shevtsova, Lilia, *Russia: Lost in Transition*, Washington, D.C.: Carnegie Endowment for International Peace, 2007.

Shlapak, David A., and Michael W. Johnson, *Reinforcing Deterrence on NATO's Eastern Flank: Wargaming the Defense of the Baltics*, Santa Monica, Calif.: RAND Corporation, RR-1253-A, June 2015. As of April 2016: http://www.rand.org/pubs/research_reports/RR1253.html

Simmons, Katie, Bruce Stokes, and Jacob Poushter, "NATO Publics Blame Russia for Ukrainian Crisis, but Reluctant to Provide Military Aid," Pew Research Center, June 10, 2015. As of March 24, 2016: http://www.pewglobal.org/2015/06/10/1-nato-public-opinion-wary-of-russia-leary-of-action-on-ukraine/

Simon, Zoltan, "Hungary Radical Party Wins By-Election in Breakthrough Vote," Bloomberg, April 12, 2015. As of March 18, 2016: http://www.bloomberg.com/news/articles/2015-04-12/hungarian-radical-party-wins-by-election-in-breakthrough-vote

Soloyov, Vladimir, "*Miroporyadok* [World Order]," YouTube, streamed live December 20, 2015. As of March 2016: https://www.youtube.com/watch?v=ZNhYzYUo42g

Sorrells, William T., Glen R. Downing, Paul J. Blakesley, David W. Pendall, Jason K. Walk, and Richard D. Wallwork, *Systemic Operational Design: An Introduction*, Fort Leavenworth, Kan.: School of Advanced Military Studies, AY 04-05, May 26, 2005.

Speaker's Advisory Group on Russia, U.S. House of Representatives, "From Friendship to Cold Peace: The Decline of the U.S. Russia Relations During the 1990s," *Russia's Road to Corruption: How the Clinton Administration Exported Government Instead of Free Enterprise and Failed the Russian People*, Washington, D.C.: U.S. Congress, via Federation of American Scientists website, September 2000. As of October 26, 2016: http://web.archive.org/web/20160415155333/http://fas.org/news/russia/2000/russia/part10.htm

Spriņģe, Inga, Donata Motuzaite, and Gunita Gailāne, "Money from Russia: Spreading Democracy in Latvia, Kremlin Style," *Re: Baltica*, March 19, 2012. As of April 2016: http://www.rebaltica.lv/en/investigations/money_from_russia/a/606/spreading_democracy_in_latvia_kremlin_style.html

Stachniak, Cezary, Twitter post, January 30, 2018. As of June 28, 2018: https://twitter.com/cezarysta/status/958419005391425536

Stamouli, Nektaria, "Russian President Vladimir Putin Aims to Renew Ties During Visit to Greece," *Wall Street Journal*, May 27, 2016. As of June 2016: http://www.wsj.com/articles/russian-president-vladimir-putin-aims-to-renew-greece-ties-during-visit-1464344808

Starikov, Nikolai, "Почему эстонцы ведут себя так нагло, а Россия так сдержанно? [Why Are Estonians Behaving So Boldly and Russia So Timidly?]," *Internet vs. TV Screen*, undated. As of February 2016:
http://www.contrtv.ru/print/2284/

Statistical Office of Estonia, Central Statistical Bureau of Latvia, and Statistics Lithuania, *2011 Population and Housing Censuses in Estonia, Latvia, and Lithuania*, 2015. As of June 1, 2016:
http://www.stat.ee/dokumendid/220923

Statistics Estonia, "Statistical Database," undated. As of March 2016:
http://pub.stat.ee/px-web.2001/Dialog/statfile1.asp

"States Hosting Expanded NATO Forces Reduce Own Level of Security—Top Russian Diplomat," RT, December 30, 2017. As of March 1, 2018:
https://www.rt.com/news/414633-nato-forces-reduce-security/

Stavljanin, Dragan, and Ron Synovitz, "'Damn Lies, Deep Crisis' in Russian Economy, Says Former Central Banker," Radio Free Europe/Radio Liberty, February 16, 2016. As of March 29, 2016:
http://www.rferl.org/content/damn-lies-deep-crisis-russia-central-banker-economy/27550209.html

"Stirring the Pot," *The Economist*, May 3, 2015. As of April 2016:
http://www.economist.com/news/europe/21645522-leader-ethnic-polish-party-tries-broaden-his-appeal-reaching-out-ethnic

Stokes, Bruce, Russia, *Putin Held in Low Regard Around the World: Russia's Image Trails the US Across All Regions*, Pew Research Center, August 5, 2015. As of March 24, 2016:
http://www.pewglobal.org/files/2015/08/Pew-Research-Center-Russia-Image-Report-FINAL-August-5-2015.pdf

Stoltenberg, Jens, *The Secretary General's Annual Report:2015*, Brussels: NATO, 2016. As of September 2, 2016:
http://www.nato.int/nato_static_fl2014/assets/pdf/pdf_2016_01/20160128_SG_AnnualReport_2015_en.pdf

Strzelecki, Marek, "Poland Opens LNG Terminal, Pledges to End Russian Dependence," Bloomberg, October 12, 2015. As of March 31, 2016:
http://www.bloomberg.com/news/articles/2015-10-12/poland-opens-lng-terminal-pledges-to-end-russian-gas-dependence

Szpala, Marta, "Russia in Serbia—Soft Power and Hard Interests," *OSW*, October 29, 2014. As of July 18, 2016:
http://www.osw.waw.pl/en/publikacje/osw-commentary/2014-10-29/russia-serbia-soft-power-and-hard-interests

Tabakova, Vesela, "Media Landscapes: Bulgaria," Maastricht, The Netherlands: European Journalism Centre, undated. As of July 14, 2016:
http://ejc.net/media_landscapes/bulgaria

Tan, Michelle, "Army Wants to Double Tanks, Boost Soldiers in Europe," *Army Times*, July 15, 2015. As of March 10, 2016:
http://www.armytimes.com/story/military/careers/army/2015/07/15/army-plans-double-equipment-boost-soldiers-europe/30187329/

Tanjug, "Russia Abstains During Vote to Extend EUFOR Mandate," B92, November 12, 2014. As of July 2016:
http://www.b92.net/eng/news/region.php?yyyy=2014&mm=11&dd=12&nav_id=92209

Taylor, Guy, "Russia Propaganda Machine Gains on U.S.," *Washington Times*, December 27, 2015. As of March 30, 2016:
http://www.washingtontimes.com/news/2015/dec/27/russia-propaganda-machine-gains-on-us/?page=all

Thomas, Timothy H., "Russia's Information Warfare Strategy: Can the Nation Cope in Future Conflicts?" *Journal of Slavic Military Studies*, Vol. 27, No. 1, 2014, pp. 101–130.

Tikk, Eneken, Kadri Kaska, and Liis Vihul, *International Cyber Incidents: Legal Considerations*, Talinn, Estonia: Cooperative Cyber Defence Centre of Excellence, 2010.

Toal, Gerard, and Carl T. Dahlman, *Bosnia Remade: Ethnic Cleansing and Its Reversal*, New York: Oxford University Press, 2011.

Toomet, Ott, "Learn English, Not the Local Language! Ethnic Russians in the Baltic States," *American Economic Review*, Vol. 101, No. 3, 2011, pp. 526–531.

Treisman, Daniel, "Why Putin Took Crimea: The Gambler in the Kremlin," *Foreign Affairs*, May–June 2016. As of July 2016:
https://www.foreignaffairs.com/articles/ukraine/2016-04-18/why-putin-took-crimea

Tremlett, Giles, "The Podemos Revolution: How a Small Group of Radical Academics Changed European Politics," *The Guardian*, March 31, 2015. As March 22, 2016:
http://www.theguardian.com/world/2015/mar/31/podemos-revolution-radical-academics-changed-european-politics

Trenin, Dmitri, *Post-Imperium: A Eurasian Story*, Washington, D.C.: Carnegie Endowment for International Peace, 2011.

"Turkey Bans Bulgarian Politicians from Entering Country—Reports," *Sofia Globe,* February 11, 2016. As of July 18, 2016:
http://sofiaglobe.com/2016/02/11/turkey-bans-bulgarian-politicians-from-entering-country-reports/

"Two-Thirds 'Loyal' to Latvia in Minority Poll," Latvian Public Broadcasting English-Language Service, August 26, 2014. As of June 1, 2016:
http://www.lsm.lv/en/article/societ/society/two-thirds-loyal-to-latvia-in-poll.a96030/

Union of Stateless People of Estonia, homepage, undated. As of March 2016: http://aliens.ee/?lang=en

United Nations High Commissioner on Refugees, *Asylum Trends in 2014: Levels and Trends in Industrialized Countries*, Geneva, Switzerland, 2014. As of March 15, 2016: http://www.unhcr.org/551128679.pdf

U.S. Army Europe, "US Army Europe to Increase Presence Across Eastern Europe," November 4, 2016. As of June 26, 2018: https://www.army.mil/article/177819/us_army_europe_to_increase_presence_across_eastern_europe

Vaksberg, Tatiana, and Alexander Andreev, "Recalling the Fate of Bulgaria's Turkish Minority," *Deutsche Welle*, December 24, 2014. As of July 2016: http://www.dw.com/en/recalling-the-fate-of-bulgarias-turkish-minority/a-18149416

Valentini, Fabio Benedetti, "Le Pen Seeks to Revive France's National Front with Name Change," Bloomberg Politics, January 7, 2018. As of January 9, 2018: https://www.bloomberg.com/news/articles/2018-01-07/le-pen-seeks-to-revive-national-france-s-front-with-name-change

van Herpen, Marcel, *Putin's Propaganda Machine: Soft Power and Russian Foreign Policy*, Lanham, Md.: Rowman and Littlefield, 2016.

Vasovic, Aleksandar, "With Russia as an Ally, Serbia Edges Toward NATO," Reuters, July 3, 2016. As of July 18, 2016: http://www.reuters.com/article/us-serbia-nato-idUSKCN0ZJ06S

"Veitman Sentenced to 15 Years for Spying for Russia," Estonian Public Broadcasting, October 30, 2013. As of April 2016: http://news.err.ee/v/news/politics/b447d2aa-07a0-4d36-82d9-44310784cbba/veitman-sentenced-to-15-years-for-spying-for-russia

Warrell, Helen, "UK Sees Surge in Wealthy Russians with Fast-Track Entries," *Financial Times*, July 31, 2014. As of March 15, 2016: http://www.ft.com/cms/s/0/df9c889a-18c6-11e4-80da-00144feabdc0.html#ixzz42WIq1KIx

Weir, Fred, "Russia's Flights Over Europe: How Much Bark, How Much Bite?" *Christian Science Monitor*, October 30, 2014. As of March 31, 2016: http://www.csmonitor.com/World/Europe/2014/1030/Russia-s-flights-over-Europe-How-much-bark-how-much-bite

Weiss, Michael, "The Estonian Spymasters," *Foreign Affairs*, June 3, 2014.

"What Does the Left Party Want for Europe?" *The Local*, May 14, 2014. As March 30, 2016: http://www.thelocal.de/20140514/what-do-die-linke-want-for-europe

White House, "Remarks by President Obama to the People of Estonia," September 3, 2014. As of March 2016:
https://www.whitehouse.gov/the-press-office/2014/09/03/remarks-president-obama-people-estonia

———, "Fact Sheet: The United States and Estonia, Latvia, and Lithuania—NATO Allies and Global Partners," August 23, 2016. As of June 26, 2018:
https://obamawhitehouse.archives.gov/the-press-office/2016/08/23/fact-sheet-united-states-and-estonia-latvia-and-lithuania-—-nato-allics

White House, *National Security Strategy of the United States of America*, December 2017. As of March 1, 2018:
https://www.whitehouse.gov/wp-content/uploads/2017/12/NSS-Final-12-18-2017-0905-2.pdf

Whitlock, Craig, and Peter Finn, "Schroeder Accepts Russian Pipeline Job," *Washington Post*, December 10, 2005. As of March 11, 2016:
http://www.washingtonpost.com/wp-dyn/content/article/2005/12/09/AR2005120901755.html

Whitmore, Brian, "Organized Crime Is Now a Major Element of Russia Statecraft," *Business Insider*, October 27, 2015. As of June 1, 2016:
http://www.businessinsider.com/organized-crime-is-now-a-major-element-of-russia-statecraft-2015-10

Wilkening, Dean, "Does Missile Defence in Europe Threaten Russia?" *Survival*, Vol. 54, No. 1, 2012, pp. 31–52.

Williams, Carol J., "Russia and Greece Consider Collaborating to Circumvent Western Sanctions," *Los Angeles Times*, June 21, 2015. As of March 21, 2016:
http://www.latimes.com/world/europe/la-fg-greece-russia-europe-20150621-story.html

Willsher, Kim, "France's Far-Right National Front Basks in Election Victory," *Los Angeles Times*, May 27, 2014. As of March 18, 2016:
http://www.latimes.com/world/europe/la-fg-france-national-front-election-20140527-story.html

Winnerstig, Mike, *Tools of Destabilization: Russian Soft Power and Non-Military in the Baltic States*, Swedish Defence Research Agency, December 2014.

World Bank, "GINI Index (World Bank Estimate)," undated. As of March 2016:
http://data.worldbank.org/indicator/SI.POV.GINI

———, "United States," undated-b. As of March 29, 2016:
http://data.worldbank.org/country/united-states

Xil, "Russians in Baltic States, 2011," licensed under CC BY-SA 3.0 via Commons, uploaded February 19, 2015. As of April 2016:
https://commons.wikimedia.org/wiki/File:Russians_in_Baltic_States_(2011).svg#/media/File:Russians_in_Baltic_States_(2011).svg

Yourish, Karen, Tim Wallace, Derek Watkins, and Tom Giratikanon, "Brussels Is Latest Target in Islamic State's Assault on West," *New York Times*, March 25, 2016. As March 30, 2016:
http://www.nytimes.com/interactive/2016/03/25/world/map-isis-attacks-around-the-world.html

Zavadski, Katie, "Putin's Propaganda TV Lies About Its Popularity," *Daily Beast*, September 17, 2015. As of March 23, 2016:
http://www.thedailybeast.com/articles/2015/09/17/putin-s-propaganda-tv-lies-about-ratings.html

Zevelev, Igor, *NATO's Enlargement and Russian Perceptions of Eurasian Political Frontiers*, Garmisch-Partenkirchen, Germany: George Marshall European Center for Security Studies, NATO, undated, p. 17. As of March 2016:
http://www.nato.int/acad/fellow/98-00/zevelev.pdf

———, "Russia's Policy Towards Compatriots in the Former Soviet Union," *Russia in Global Affairs*, Vol. 6, No. 1, January–March 2008.

Zuvela, Maja, "Bosnia to Investigate Suspected Serb Paramilitary Group," Reuters, January 16, 2018. As of February 27, 2018:
https://www.reuters.com/article/us-bosnia-security-paramilitary/bosnia-to-investigate-suspected-serb-paramilitary-group-idUSKBN1F51ZW